PROGRESS IN CLINICAL AND BIOLOGICAL RESEARCH

RECENT TITLES

Vol 269: **EORTC Genitourinary Group Monograph 5: Progress and Controversies in Oncological Urology II,** Fritz H. Schröder, Jan G.M. Klijn, Karl H. Kurth, Herbert M. Pinedo, Ted A.W. Splinter, Herman J. de Voogt, *Editors*

Vol 270: **Cell-Free Analysis of Membrane Traffic,** D. James Morré, Kathryn E. Howell, Geoffrey M.W. Cook, W. Howard Evans, *Editors*

Vol 271: **Advances in Neuroblastoma Research 2,** Audrey E. Evans, Giulio J. D'Angio, Alfred G. Knudson, Robert C. Seeger, *Editors*

Vol 272: **Bacterial Endotoxins: Pathophysiological Effects, Clinical Significance, and Pharmacological Control,** Jack Levin, Harry R. Büller, Jan W. ten Cate, Sander J.H. van Deventer, Augueste Sturk, *Editors*

Vol 273: **The Ion Pumps: Structure, Function, and Regulation,** Wilfred D. Stein, *Editor*

Vol 274: **Oxidases and Related Redox Systems:** Tsoo E. King, Howard S. Mason, Martin Morrison, *Editors*

Vol 275: **Electrophysiology of the Sinoatrial and Atrioventricular Nodes,** Todor N. Mazgalev, Leonard S. Dreifus, Eric L. Michelson, *Editors*

Vol 276: **Prediction of Response to Cancer Therapy,** Thomas C. Hall, *Editor*

Vol 277: **Advances in Urologic Oncology,** Nasser Javadpour, Gerald P. Murphy, *Editors*

Vol 278: **Advances in Cancer Control: Cancer Control Research and the Emergence of the Oncology Product Line,** Paul F. Engstrom, Paul N. Anderson, Lee E. Mortenson, *Editors*

Vol 279: **Basic and Clinical Perspectives of Colorectal Polyps and Cancer,** Glenn Steele, Jr., Randall W. Burt, Sidney J. Winawer, James P. Karr, *Editors*

Vol 280: **Plant Flavonoids in Biology and Medicine II: Biochemical, Cellular, and Medicinal Properties,** Vivian Cody, Elliott Middleton, Jr., Jeffrey B. Harborne, Alain Beretz, *Editors*

Vol 281: **Transplacental Effects on Fetal Health,** Dante G. Scarpelli, George Migaki, *Editors*

Vol 282: **Biological Membranes: Aberrations in Membrane Structure and Function,** Manfred L. Karnovsky, Alexander Leaf, Liana C. Bolis, *Editors*

Vol 283: **Platelet Membrane Receptors: Molecular Biology, Immunology, Biochemistry, and Pathology,** G.A. Jamieson, *Editor*

Vol 284: **Cellular Factors in Development and Differentiation: Embryos, Teratocarcinomas, and Differentiated Tissues,** Stephen E. Harris, Per-Erik Mansson, *Editors*

Vol 285: **Non-Radiometric Assays: Technology and Application in Polypeptide and Steroid Hormone Detection,** Barry D. Albertson, Florence P. Haseltine, *Editors*

Vol 286: **Molecular and Cellular Mechanisms of Septic Shock,** Bryan L. Roth, Thor B. Nielsen, Adam E. McKee, *Editors*

Vol 287: **Dietary Restriction and Aging,** David L. Snyder, *Editor*

Vol 288: **Immunity to Cancer. II,** Malcolm S. Mitchell, *Editor*

Vol 289: **Computer-Assisted Modeling of Receptor–Ligand Interactions: Theoretical Aspects and Applications to Drug Design,** Robert Rein, Amram Golombek, *Editors*

Vol 290: **Enzymology and Molecular Biology of Carbonyl Metabolism 2: Aldehyde Dehydrogenase, Alcohol Dehydrogenase, and Aldo-Keto Reductase,** Henry Weiner, T. Geoffrey Flynn, *Editors*

Vol 291: **QSAR: Quantitative Structure-Activity Relationships in Drug Design,** J.L. Fauchère, *Editor*

Vol 292: **Biological and Synthetic Membranes,** D. Allan Butterfield, *Editor*

Vol 293: **Advances in Cancer Control: Innovations and Research,** Paul N. Anderson, Paul F. Engstrom, Lee E. Mortenson, *Editors*

Vol 294: **Development of Preimplantation Embryos and Their Environment,** Koji Yoshinaga, Takahide Mori, *Editors*

Vol 295: **Cells and Tissues: A Three-Dimensional Approach by Modern Techniques in Microscopy,** Pietro M. Motta, *Editor*

Vol 296: **Developments in Ultrastructure of Reproduction,** Pietro M. Motta, *Editor*

Vol 297: **Biochemistry of the Acute Allergic Reactions: Fifth International Symposium,** Bruce U. Wintroub, Alfred I. Tauber, Arlene Stolper Simon, *Editors*

Vol 298: **Skin Carcinogenesis: Mechanisms and Human Relevance,** Thomas J. Slaga, Andre J.P. Klein-Szanto, R.K. Boutwell, Donald E. Stevenson, Hugh L. Spitzer, Bob D'Motto, *Editors*

Vol 299: **Perspectives in Shock Research: Metabolism, Immunology, Mediators, and Models,** John C. Passmore, *Editor,* Sherwood M. Reichard, David G. Reynolds, Daniel L. Traber, *Co-Editors*

Vol 300: **Alpha$_1$-Acid Glycoprotein: Genetics, Biochemistry, Physiological Functions, and Pharmacology,** Pierre Baumann, Chin Bin Eap, Walter E. Müller, Jean-Paul Tillement, *Editors*

Vol 301: **Prostaglandins in Clinical Research: Cardiovascular System,** Karsten Schrör, Helmut Sinzinger, *Editors*

Vol 302: **Sperm Measures and Reproductive Success: Institute for Health Policy Analysis Forum on Science, Health, and Environmental Risk Assessment,** Edward J. Burger, Jr., Robert G. Tardiff, Anthony R. Scialli, Harold Zenick, *Editors*

Vol 303: **Therapeutic Progress in Urological Cancers,** Gerald P. Murphy, Saad Khoury, *Editors*

Vol 304: **The Maillard Reaction in Aging, Diabetes, and Nutrition,** John W. Baynes, Vincent M. Monnier, *Editors*

Vol 305: **Genetics of Kidney Disorders,** Christos S. Bartsocas, *Editor*

Vol 306: **Genetics of Neuromuscular Disorders,** Christos S. Bartsocas, *Editor*

Vol 307: **Recent Advances in Avian Immunology Research,** Balbir S. Bhogal, Guus Koch, *Editors*

Vol 308: **Second Vienna Shock Forum,** Günther Schlag, Heinz Redl, *Editors*

Vol 309: **Advances and Controversies in Thalassemia Therapy: Bone Marrow Transplantation and Other Approaches,** C. Dean Buckner, Robert Peter Gale, Guido Lucarelli, *Editors*

Vol 310: **EORTC Genitourinary Group Monograph 6: BCG in Superficial Bladder Cancer,** Frans M.J. Debruyne, Louis Denis, Ad P.M. van der Meijden, *Editors*

Vol 311: **Molecular and Cytogenetic Studies of Non-Disjunction,** Terry J. Hassold, Charles J. Epstein, *Editors*

Vol 312: **The Ocular Effects of Prostaglandins and Other Eicosanoids,** Laszlo Z. Bito, Johan Stjernschantz, *Editors*

Vol 313: **Malaria and the Red Cell 2,** John W. Eaton, Steven R. Meshnick, George J. Brewer, *Editors*

Vol 314: **Inherited and Environmentally Induced Retinal Degenerations,** Matthew M. LaVail, Robert E. Anderson, Joe G. Hollyfield, *Editors*

Vol 315: **Muscle Energetics,** Richard J. Paul, Gijs Elzinga, Kazuhiro Yamada, *Editors*

Please contact publisher for information about previous titles in this series.

MOLECULAR AND CYTOGENETIC STUDIES OF NON-DISJUNCTION

MOLECULAR AND CYTOGENETIC STUDIES OF NON-DISJUNCTION

Proceedings of the Fifth Annual National Down Syndrome Society Symposium, Held in New York, NY, December 1–2, 1988

Editors

Terry J. Hassold

Department of Pediatrics
Emory University Medical School
Atlanta, Georgia

Charles J. Epstein

Department of Pediatrics
University of California
San Francisco, California

ALAN R. LISS, INC. • NEW YORK

Address all Inquiries to the Publisher
Alan R. Liss, Inc., 41 East 11th Street, New York, NY 10003

Printed in United States of America

The publication of this volume was facilitated by the authors and editors who submitted the text in a form suitable for direct reproduction without subsequent editing or proofreading by the publisher.

Library of Congress Cataloging-in-Publication Data

National Down Syndrome Society (U.S.). Symposium (5th : 1988 : New York, N.Y.)
Molecular and cytogenetic studies of non-disjunciton : proceedings of the Fifth Annual National Down Syndrome Society Symposium held in New York, N.Y., December 1–2, 1988 / editors, Terry J. Hassold, Charles J. Epstein.
p. cm. — (Progress in clinical and biological research ; v. 311)
Includes bibliographies and index.
ISBN 0-8451-5161-4
1. Down's syndrome—Etiology—Congresses. 2. Cytogenetics—Congresses. 3. Pathology. Molecular—Congresses. I. Hassold, Terry J. II. Epstein, Charles J. III. Title. IV. Series.
RC571.N35 1988
616.85′8842042—dc20

89-2709
CIP

Contents

Contributors

Stylianos E. Antonarakis, Genetics Unit, Department of Pediatrics, The Johns Hopkins University School of Medicine, Baltimore, MD 21205 **[29,45,81]**

Monika Buraczynska, Howard Hughes Medical Institute, and Department of Pediatrics, University of Michigan, Ann Arbor, MI 48109 **[101]**

Aravinda Chakravarti, Human Genetics Division, University of Pittsburgh, Pittsburgh, PA 15261 **[29,45]**

Maimon M. Cohen, Division of Human Genetics, Departments of Obstetrics/Gynecology and Pediatrics, University of Maryland Medical School, Baltimore, MD 21201 **[81]**

Ranjan Deka, Human Genetics Division, University of Pittsburgh, Pittsburgh, PA 15261 **[45]**

Ursula Eichenlaub-Ritter, Department of Biology, Gentechnology, University of Bielefeld, Bielefeld, FRG **[199]**

JoAnn Engebrecht, Department of Biology, Yale University, New Haven, CT 06511 **[303]**

Anna Estop, Department of Medical Genetics, West Penn Hospital, Pittsburgh, PA 15224 **[221]**

Robert E. Ferrell, Human Genetics Division, University of Pittsburgh, Pittsburgh, PA 15261 **[45]**

Viola Freeman, Department of Pediatrics, McMaster University, Hamilton, Ontario, Canada L8S 3Z5 **[101]**

Meg Grantham, Department of Pediatrics, Emory University, Atlanta, GA 30322 **[101]**

John Harvey, Wessex Regional Cytogenetic Unit S.G.I., Salisbury, Wilts, U.K. SP2 7SX **[135]**

Terry J. Hassold, Department of Pediatrics, Division of Medical Genetics, Emory University Medical School, Atlanta, GA 30322 **[101,115,135]**

R. Scott Hawley, Department of Molecular Genetics, Albert Einstein College of Medicine, Bronx, NY 10461 **[277]**

Philip A. Hieter, Department of Molecular Biology and Genetics, The Johns Hopkins University School of Medicine, Baltimore, MD 21205 **[29]**

The numbers in brackets are the opening page numbers of the contributors' articles.

Ernest B. Hook, School of Public Health, Department of Social and Administrative Health Sciences, University of California, Berkeley, CA 94720 **[1]**

Patricia Jacobs, Wessex Regional Cytogenetic Unit S.G.I., Salisbury, Wilts, U.K. SP2 7SX **[135]**

Dagmar K. Kalousek, Department of Pathology, University of British Columbia, Vancouver, B.C., Canada **[153]**

David M. Kurnit, Howard Hughes Medical Institute, and Department of Human Genetics, University of Michigan, Ann Arbor, MI 48109 **[101]**

John G. Lewis, Genetics Unit, Department of Pediatrics, The Johns Hopkins University School of Medicine, Baltimore, MD 21205 **[29,81]**

Dan L. Lindsley, Department of Biology, University of California, San Diego, La Jolla, CA 92092 **[263]**

Partha P. Majumder, Human Genetics Division, University of Pittsburgh, Pittsburgh, PA 15261 **[45]**

Kristen May, Department of Pediatrics, Emory University School of Medicine, Atlanta, GA 30322 **[135]**

Mary Kay McCormick, Genetics Unit, Department of Pediatrics, The Johns Hopkins University School of Medicine, Baltimore, MD 21205 **[29]**

Thomas M. Menees, Department of Biology, Yale University, New Haven, CT 06511 **[303]**

Margareta Mikkelsen, Department of Medical Genetics, The J.F. Kennedy Institute, Glostrup, Denmark DK-2600 **[183]**

Hanne Poulsen, Department of Medical Genetics, The J.F. Kennedy Institute, Glostrup, Denmark DK-2600 **[183]**

Beth M. Rockmill, Department of Biology, Yale University, New Haven, CT 06511 **[303]**

G. Shirleen Roeder, Department of Biology, Yale University, New Haven, CT 06511 **[303]**

Diane Roulston, Division of Human Genetics, Departments of Obstetrics/Gynecology and Pediatrics, University of Maryland School of Medicine, Baltimore, MD 21201 **[81]**

Stuart Schwartz, Division of Human Genetics, Departments of Obstetrics/Gynecology and Pediatrics, University of Maryland Medical School, Baltimore, MD 21201 **[81]**

Stephanie Sherman, Department of Pediatrics, Emory University, Atlanta, GA 30322 **[101]**

Susan A. Slaugenhaupt, Human Genetics Division, University of Pittsburgh, Pittsburgh, PA 15261 **[45]**

Gordon D. Stewart, Howard Hughes Medical Institute, and Department of Human Genetics, University of Michigan, Ann Arbor, MI 48109 **[101]**

Richard T. Surosky, Department of Biochemistry, Cell and Molecular Biology, Cornell University, Ithaca, NY 14853; present address: Department of Molecular Genetics, University of Chicago, Chicago, IL 60637 **[327]**

Urvashi Surti, Department of Pathology, University of Pittsburgh, Pittsburgh, PA 15261 **[45]**

Norma Takaesu, Department of Pediatrics, Division of Medical Genetics, Emory University Medical School, Atlanta, GA 30322 **[115]**

Emily A. Thompson, Department of Biology, Yale University, New Haven, CT 06511 **[303]**

Niels Tommerup, Department of Medical Genetics, The J.F. Kennedy Institute, Glostrup, Denmark DK–2600 **[183]**

Bik-Kwoon Tye, Department of Biochemistry, Cell and Molecular Biology, Cornell University, Ithaca, NY 14853 **[327]**

Irene Uchida, Department of Pediatrics, McMaster University, Hamilton, Ontario, Canada L8S 3Z5 **[101]**

M. Vekemans, Department of Pediatric Pathology, The Montreal Children's Hospital, and Centre for Human Genetics, McGill University, Montreal, Canada **[245]**

Dorothy Warburton, Department of Pediatrics and Department of Genetics and Development, College of Physicians and Surgeons, Columbia University, New York, NY 10032 **[165]**

Andrew C. Warren, Genetics Unit, Department of Pediatrics, The Johns Hopkins University School of Medicine, Baltimore, MD 21205 **[29,45]**

Preface

In 1959, Jerome Lejeune and his colleagues described the presence of an extra chromosome 21 in Down syndrome, thereby providing the first evidence of an association between chromosome imbalance and human disease. Thirty years have passed since this landmark discovery. Human cytogenetics is now an integral part of diagnostic medicine, and numerical and structural abnormalities involving all human chromosomes have been identified and characterized. Furthermore, Down syndrome research has flourished: cytogenetic studies of the 1960s and 1970s led to the identification of Down syndrome individuals with translocations, partial trisomy 21, and mosaic trisomy 21, and, more recently, cytogenetic and molecular studies have begun to elucidate the structure of chromosome 21 and to pinpoint genetic loci important in the development of the Down syndrome phenotype. These and related aspects of Down syndrome have been the subject of the four earlier scientific symposia sponsored by the National Down Syndrome Society and published as *Molecular Structure of the Number 21 Chromosome and Down Syndrome,* G.F. Smith (ed.), *Annals of the New York Academy of Sciences,* Volume 450, 1985; *The Neurobiology of Down Syndome,* C.J. Epstein (ed.) (Raven Press, New York, 1986); *Oncology and Immunology of Down Syndrome,* E.E. McCoy and C.J. Epstein (eds.) (Alan R. Liss, New York, 1987); and *The Psychobiology of Down Syndrome,* L. Nadel (ed.) (MIT Press, Cambridge, 1988).

Despite the great progress that has been made in the areas covered by these volumes, surprisingly little is yet known about the origin of Down syndrome in terms of the underlying mechanisms responsible for the additional chromosome 21. Cytogenetic studies have shown that the extra chromosome is more often maternal than paternal in origin and that it most commonly derives from an error at the first meiotic division in the mother. However, these studies have provided little insight into the causes of the meiotic errors and have not provided an explanation for the well-known association between increasing maternal age and trisomy.

The fifth annual National Down Syndrome Society Symposium, held on December 1–2, 1988, was devoted to a discussion of non-disjunction and brought together population geneticists, cytogeneticists, molecular biologists, and embryologists studying this process in humans and in experimental animals. The meeting was organized to highlight two topics: the association between meiotic recombination and chromosome disjunction in different organisms and the basis of the relationship between increasing maternal age and non-disjunction in humans. Following introductory chapters on the impact of trisomy 21 in humans and the molecular organization of chromosome 21, the first half of the book describes recent molecular and cytogenetic studies of non-disjunction in humans. The second half of the book is devoted to analyses of non-disjunction in the mouse, *Drosophila,* and yeast and is presented in the belief that an eventual understanding of non-disjunction of human chromosome 21 will depend on studies of these and other species in addition to studies of humans.

This volume is directed to several groups, including clinicians caring for individuals with Down syndrome or other trisomic conditions, geneticists studying chromosome behavior in man and other species, and researchers interested in chromosome organization. It is our hope that the material presented in this book will stimulate additional research into the origin and nature of trisomy, the most common genetic disorder in humans.

Terry J. Hassold
Charles J. Epstein

Acknowledgments

The Symposium on Molecular and Cytogenetic Studies of Non-disjunction was sponsored by The National Down Syndrome Society, under the leadership of Elizabeth Goodwin, President. The organization of the symposium was expertly handled by Donna M. Rosenthal, Executive Director of the Society, without whose efforts neither the conference nor this volume would have been possible. Both the symposium and this volume were made a reality by the generous support of the Ambrose Monell Foundation, Republic New York Corporation, St. Moritz Hotel, and British Airways.

Additional information about the work of the National Down Syndrome Society can be obtained from the Society at 141 Fifth Avenue, New York, New York 10010. The telephone numbers of the Society are (212) 460-9330 and (800) 221-4602 (outside New York State).

Molecular and Cytogenetic Studies of
Non-Disjunction, pages 1–27

ISSUES PERTAINING TO THE IMPACT AND ETIOLOGY OF TRISOMY 21 AND OTHER ANEUPLOIDY IN HUMANS; A CONSIDERATION OF EVOLUTIONARY IMPLICATIONS, MATERNAL AGE MECHANISMS, AND OTHER MATTERS

Ernest B. Hook

School of Public Health, Department of Social and Administrative Health Sciences, University of California, Berkeley CA 94720

One attraction of a symposium such as this is that it provides a forum for interactive presentation of ideas, concepts and hypotheses still in a preliminary stage. It also allows us to consider speculatively what approaches would be useful to extend existing knowledge. Below I consider a number of suggestions and ideas which I believe warrant further consideration. Before doing this, I note briefly some aspects of: i) the historical recognition of Down syndrome; and ii) the impact of aneuploidy.

HISTORICAL ASPECTS

The early history of Down syndrome has been reviewed in a fascinating article by Volpe, 1986. As we search for the eventual answers to the etiology of this disorder it is worthwhile to recall the gradual evolution of clinical knowledge in the field. Volpe asked why the syndrome was not recognized prior to Down's 1866 report and concluded that it was probably because of the confusion with "cretinism" which was, earlier, a general term for a variety of mental deficiencies. In the 19th century for instance, even after Down's publication, it was some years before the condition was widely recognized as distinctive. The report in 1900 by Sutherland that Down syndrome cases do not respond to thyroid but cretins do was a major step in establishment of the syndrome, according to Volpe.

A good deal of subsequent history, useful clinical details, and descriptive analyses on Down syndrome are presented by Penrose and Smith, 1966 (the second edition of which is by Smith and Berg, 1976).

IMPACT OF ANEUPLOIDY

The impact of aneuploidy upon our species has been reviewed extensively elsewhere (Hook, 1985a). I summarize the main points in table 1. Spontaneous embryonic or fetal death in recognized pregnancies is the greatest biological consequence of aneuploidy. About 30% of all spontaneous abortions in recognized pregnancies have a chromosome abnormality (Hook, 1981a). Almost all are numerical abnormalities, although there is a small proportion, about 1-3%, attributable to structural abnormality detectable with low level banding. (The value of 30% incidentally, predicts a rate of all recognized conceptuses with chromosome abnormality of about 5%. This follows from the formula $z = qx + y(1-q)$ where z is the overall rate, x is the proportion of (recognized) abortuses with chromosome abnormalities, about .30, q is the proportion of recognized conceptuses aborted, about .15, and p is the proportion of live births with abnormality, about .006.) The toll of aneuploidy may be even greater if we consider unrecognized pregnancies, those that terminate before five weeks gestational age (from the first day of the last menstrual period). Some indirect evidence on this point follows from reports of Martin (1985), Brandriff et al (1986) and Zenzes (1987) on aneuploidy in sperm, Schmiady et al (1987) and Wramsby et al (1987) on oocytes, and Angell et al (1986) on early embryos.

Effects of aneuploidy often are not appreciated, even by clinicians who work with affected cases. For example, some years ago (Hook, 1983) I estimated that Down syndrome comprised about 6% to 7% of all live birth infants with congenital heart defect. If one adds in Patau syndrome (47,+13), Edwards syndrome (47,+18), Turner syndrome (45,X and others), and other syndromes produced by aneuploidy, then this proportion rises to about 10%. The mathematical derivation is straight forward and follows from the following type of considerations:

i) about 1/1000 live born infants have Down syndrome;

ii) about 50% of Down syndrome children have or will manifest a congenital heart defect;

iii) about 0.6% to 0.7% of all live born infants have a congenital heart defect. (This is derivable from data of Heinonen et al, 1976, p.245.)

Thus [(1/2)(1/1000)/(.007)] = 1/14 or about 7% of all infants with a heart defect have Down syndrome. By a similar derivation, other aneuploidy contributes about 3% to 4%. But pediatric cardiologists and others were incredulous when I cited this figure. They insisted they didn't see this frequency in their practices. Their response illustrates I believe, a negative ascertainment bias. Aneuploid children with congenital heart defects probably either die early or have surgical correction, and so leave less impression upon clinicians than others with a different spectrum of abnormality who are seen more frequently in clinic.

Subsequently, a survey by Ferenez et al, 1987 found about 13% of infants with congenital heart defect had a chromosome abnormality, consistent with the earlier estimate.

EVOLUTIONARY IMPLICATIONS

A major question is why aneuploidy is so prevalent in our species. Embryonic and fetal loss attributable to chromosome abnormality takes so high a toll that one wonders how the species could waste reproductive potential in this manner. It appears plausible that there is some biological advantage of aneuploidy to the species.

Three types of evolutionary hypotheses may be visualized that account for the high frequency of chromosome abnormality. One suggests that predisposing factors have selective value because aneuploidy enables appropriate spacing of pregnancy (Jacobs, 1975; see also Warburton, 1987). Death of a conceptus makes it more likely on this hypothesis that a prior infant would survive because parental attention is not diverted to a subsequent infant. I have been unable to discover, however, any evidence for a shorter interpregnancy interval before the birth of a Down syndrome individuals than before the birth of normal infants as might be implied by this theory. Also, Down syndrome infants are not less frequently the first born (after correction for maternal age). (See review in Hook, 1981b or Hook, 1982.) Both of these observations are contrary to the expectation of the hypothesis. But of course, 47,+21 (in live births) makes up only a tiny proportion of all recognized aneuploidy, perhaps 0.3%, and may not be representative of the vast bulk of aneuploidy. Moreover, pregnancy control and demographic patterns are now quite different from those at the time our species evolved. Current trends may not reflect historical patterns that were important in evolution.

A particular mystery is why, from an evolutionary perspective, the prevalence of aneuploidy is so much higher at advanced maternal ages. (For data and discussion bearing on maternal age trends, see Hook, 1985b; Warburton's chapter elsewhere in this volume; and further discussion below.) Projections from data on live births, amniocentesis studies, and spontaneous embryonic fetal death suggest that over age 40, the proportion of aneuploid (recognized) conceptuses is close to 50% (Hook, unpublished). This is at an age moreover when fertility is diminished.

I believe that the very high prevalence of aneuploid conceptuses at older ages has not been selected for in the course of evolution. Rather the older age effect may result only from the "burn out" of biological processes at ages at which early humans usually died. Early fertility of prehistoric women -- and early death -- would not have exerted as much selection pressure against reproductive accidents at older ages as at younger ones. Now that life is not as "short, nasty and brutish" as in prehistory, fertile women survive to older ages but experience an increasing failure in fidelity of chromosome transmission to offspring.

I emphasize that this postulated shorter "window" of reproduction in prehistoric women because of earlier mortality is still a hypothesis. I am now seeking direct evidence on this point from anthropologists who have estimated ages at death of prehistoric populations.

In any event, aneuploidy in offspring of younger mothers, say those under age 25 (in which chromosome abnormality may make up about 10% to 15% of all recognized embryonic and fetal loss) is more likely to have some selective significance in the evolution of our species than aneuploidy in offspring of older women. It is this group, I believe, in which the search for evidence for selective factors is likely to be most fruitful.

Another evolutionary hypothesis is that pregnancy with aneuploid embryo or fetus provides some benefit to the mother. No advantageous physiological effects consequent to pregnancy with an aneuploid fetus have yet been defined to my knowledge. But the increase in thyroid autoimmunity in women who have had aneuploid infants (see review by Cuckle et al, 1988) may be a consequence of being pregnant with the affected child and not be a predisposing factor to non-disjunction as some have thought. There may be other changes in maternal auto-immunity which are not deleterious. The changes in serum proteins of mothers pregnant with Down syndrome fetuses (e.g. alpha feto protein and chorionic gonadotrophin) illustrate another mechanism that might produce some long term maternal physiological effect.

To summarize: several different types of evolutionary explanations may be cited for the high frequency of aneuploidy:

i) an increased likelihood of survival of normal children in a sibship if an aneuploid conceptus occurs;

ii) an increased likelihood of survival of the parent of an aneuploid conceptus because of either a) pleiotropic effects of predisposing genes or b) favorable physiological effects of carrying an aneuploid conceptus (or of any spontaneously aborted conceptus for that matter);

iii) at least at older ages, an unleashing of deteriorating factors -- "a burn out" of biological factors that ensure normal euploidy -- as women survive beyond the ages at which selection occurred upon reproduction.

ETIOLOGY AND RISK FACTORS

In thinking about etiologies we are severely constricted by how few variables have been established as preconceptual risk factors for aneuploidy. For 47,+21 I know of only three: 1) maternal age; 2) the presence in the parent of a 47,+21 line; and 3) the history of a previous 47,+21 offspring. And the third may in fact be only a consequence of the second, although often this cannot be proven. (References and discussion of some other putative preconceptual risk factors not discussed further below may be found in Hook, 1984 or elsewhere in this volume in several chapters by others.)

In the rest of the discussion I will consider only aneuploidy in the more restricted sense; numerical abnormalities deviating from a multiple of the ploidy number. I will not consider translocations, except when they may influence the number of chromosomes. (For a review of some aspects of structural rearrangements see Daniel et al, 1988; Hook and Cross, 1987. The only established risk factor for this type of partial or segmental aneuploidy is the presence of related structural abnormality in one parent. It does appear likely that very high doses of parental gonadal ionizing radiation can also induce structural abnormalities, but this evidence to date is based only upon study of gametes or pregametic germinal cells, and has not been documented in the populations in Hiroshima or Nagasaki (Awa et al, 1987).)

There are of course a number of postconceptual risk factors such as low maternal serum alpha fetoprotein, elevated human chorionic gonadotrophin, delayed fetal growth, diminished embryonic or fetal viability, and certain types of malformations or phenotypic variants detectable with prenatal imaging, all of which are associated with a higher probability of 47,+21 and other aneuploidies. (The only clear exception to my knowledge is 47,XYY.) But such factors are more pertinent to differential survival of affected conceptuses than to the etiology of the abnormality.

ETHNIC AND RACIAL VARIATION

Possible ethnic and racial variation in rates of abnormalities are among the most interesting postulated preconceptual variables. Such putative variation could reflect, of course, environmental or genetic factors. There are only a few studies not biased by differential ascertainment in one of the groups -- usually rates are reported to be higher in whites than non-whites -- and/or flawed by failure to adjust for maternal age.

One study on ethnic differences in Down syndrome makes a prima facie case for higher rates in non-European Jews than European Jews in Israel (Hook and Harlap, 1978). (Rates in Jews of European origin were the same as rates reported in other European populations, but Jews from Asia or Africa had higher rates.)

In the cited study I am concerned about the theoretical possibility that women of non-European origin but not the Europeans may have reported falsely low ages. This would result in falsely higher maternal age adjusted rates in the non-Europeans and the observed ethnic differences. Dr. Harlap, who collected the data, assures me this was unlikely. The only evidence for the bias is in fact the difference in rates.

For ten years I have been trying to repeat these observations in another Israeli data set, but have been unable to find anyone to collect or provide pertinent raw data, let alone assurances that biases have not entered in their compilation.

There are two relatively good studies on white-black differences in the Down syndrome phenotype. One makes a prima facie case for a higher rate in whites at older ages (Stark and White, 1977), the other does so for blacks at older ages (Sever et al, 1970). The net effect of course is no difference.

One interesting trend is the lower rate of XYYs in black males than white males. In fact, in the consecutive new born studies in the U.S.A. there have been <u>no</u> 47,XYYs in 3,732 black male live births compared to 18 in 19,515 white male live births, a difference signifi-cant by a Fisher's exact test at the .05 level (Hook, 1984). (The XYY genotype is also proportionately less represented among blacks than whites in penal and "mental penal" settings, although blacks with XYY genotypes do occur (Hook, 1974).) If real, this difference almost certainly reflects a difference in a paternal mechanism and one acting, it is likely, on 2nd division non-disjunction. I am aware of no other trisomy for which a significant fraction can be assigned to paternal 2nd division non-disjunction. It would be worthwhile nevertheless to look at other trisomies using cytologic and biochemical markers to determine if the small proportion in which the extra chromosome is of paternal 2nd division origin exhibits a racial or ethnic difference similar to that found for the XYY.

The XXY karyotype in new borns does not exhibit the same trend as XYY (Hook, 1974). In fact, the proportion of XXYs among new borns is slightly higher in blacks than whites, a trend that exists despite the higher average maternal age in whites in the general population. (Age adjustment can't be made because of lack of precise data on maternal age in the live birth cytogenetic series.)

Beyond the obvious requirement to confirm the above findings in other populations, we need good data on all possible numerical abnormalities in as many ethnic and racial groups as possible. By good data, I mean those in which differential ascertainment has not occurred and in which maternal age is reported and adjusted for. This is important not only for purposes of etiologic investigation, but for accurate genetic counseling. I hope in the next few years we will have more definitive data at least from amniocentesis studies.

True racial and/or ethnic differences in aneuploidy could reflect variation either in genetic predisposition to aneuploidy or in environmental influences. Either of these could act at different stages of meiosis (or even mitosis in the early conceptus). To date most of the comparisons have been very crude and no one has, to my knowledge, looked at the possibility of ethnic or racial variations in aneuploidy associated with the stage and parent of origin or of non-disjunction. For example, for Down syndrome, the few analyses of racial and ethnic difference have been based upon phenotype, most of which is caused by 47,+21, in which the extra chromosome is of maternal 1st division meiotic origin in about 80% of cases. Conceivably there might be some major ethnic and/or racial variation in the other stages (mat-II, pat-I, pat-II and mitotic) which, because of their relative scarcity were not dissected out. Ethnic and racial comparisons of trisomy would benefit from stage specific analyses of the origin of the extra chromosome. And indeed, as implied in the discussion of the XYY genotype above, there is some reason to anticipate differences among populations in trisomy in which the extra chromosome has pat-II origin.

Lastly, there is an interesting racial variation in hydatidiform moles which is more frequent in Asians, especially Filipinos, than in other races (Jacobs et al, 1982). Moles may result from meiotic failure. The karyotypes are not aneuploid, but it is possible that etiologic factors during meiosis related to those that produce aneuploidy are involved.

INBREEDING AND CONSANGUINITY

More direct evidence for a specific etiology would emerge if inbred parents were found to be at high risk of having children with aneuploidy. In experimental animals various genes predisposing to aneuploidy are well known (see Baker et al, 1975 and Lindsley, elsewhere in this volume). But with one possible exception, searches for evidence of such genes in humans have been at best only suggestive. The exception is a strongly positive study of offspring of consanguineous parents in Kuwait (Alfi et al, 1980). Some regarded this report as implausible because of its apparent discrepancies from other studies on the same topic. Yet these putative negative studies elsewhere are irrelevant. After all, such genes may occur preferentially in a particular population.

It would be of interest to have not only a confirmation of the trend in another Kuwaiti population, but also data on other nearby populations in the mid-east likely to share the same gene pool. And moreover we also need a determination as to whether the differences apply to all "stages" of origin or just one.

The older evidence is moreover, not as negative as interpreted by the original authors, e.g. Matsunaga, 1966 or Forssman and Akesson, 1967. Slight or even modest consanguinity effects have not been excluded. (For extensive review and discussion of these earlier studies see Hook, 1981b or Hook, 1982).

OTHER FAMILIAL FACTORS

Theoretically, genes predisposing to non-disjunction can be classified in several different ways. One operational method, for example, would be to distinguish genes resulting in non-disjunction of a specific chromosome from genes that can produce non-disjunction of different chromosomes. Hecht, 1977 has referred to the consequences of these two different types as homoaneuploidy -- recurrence of the same abnormality -- and heteroaneuploidy -- the occurrence of different abnormalities in siblings at a rate greater than that expected by chance. The occurrence of homoaneuploidy would not prove the existence of predisposing genes because such an outcome may of course be caused by parental mosaicism as has been demonstrated occasionally. On the other hand, familial heteroaneuploidy is better evidence for genetic predisposition, although certainly recurrent environmental factors could also be invoked as a possible cause.

What sparse systematic data I have been able to find in live births and cases studied at amniocentesis provide little evidence for at least a strong heteroaneuploid effect, although a weak effect cannot be excluded. But highly suggestive evidence for heteroaneuploidy for trisomy was reported in spontaneous abortuses. (See references in Warburton et al, 1987.)

For heteroaneuploid recurrence of trisomy among abortuses, the odds that a later pregnancy is a trisomic abortus given that an earlier pregnancy is trisomic (a/b), is much higher than the odds if the earlier pregnancy is a non-trisomic spontaneous abortus (c/d). Thus the relative odds, (a/b):(c/d) or ad/bc are markedly greater than 1.0. (In this formulation a is the number of "sibships" in which the 1st and 2nd pregnancies result in trisomic spontaneous abortuses, b the number in which the 1st pregnancy is a trisomic abortus, and the 2nd a non-trisomic abortus, c is the number with a pattern opposite to b, and d the number in which neither spontaneous abortus is trisomic.) This high relative odds was interpreted originally as convincing evidence for heteroaneuploid effects.

Closer scrutiny (by Warburton et al, 1987) revealed, however, that it is most likely the relative odds (ad/bc) are high not because of an increase in the number of heteroaneuploid abortus pairs, a in the equation but because of two other factors: i) maternal ageing between the time of the 1st and 2nd offspring -- the earlier analyses did not adjust for maternal age; and ii) an increase in d, the recurrent non-trisomic spontaneous abortion pairs. Recurrent environmental or biologic factors in a woman such as smoking, "incompetent cervix", etc. which predispose to spontaneous abortions primarily of non-trisomic fetuses that otherwise would survive inflate the value of d in the calculation of the relative odds defined above.

Another type of familial aneuploidy is pertinent to an etiological mechanism suggested by the distributive pairing hypothesis of Grell (1962, 1971, 1985). This is the occurrence of unrelated chromosome abnormalities in offspring of parents carrying abnormalities which do not inhibit fertility. A 47,+21 offspring born to a parent with a (non-21) small extra structurally abnormal chromosome or a balanced structural rearrangement is an example. Despite anecdotal case reports (see references and discussion in Grell, 1971), unbiased data in humans from amniocentesis studies provide evidence against a strong effect although they don't exclude a weak one (Daniel et al, 1988). Only collection of more data will allow a strict test of the effect in humans.

TIMING OF FERTILIZATION

Some etiological theories suggest that production of a 47,+21 or other aneuploid conceptus is related to various biologic factors around the time of fertilization, for example, some factor affecting the relationship of timing of ovulation to timing of fertilization. A provocative hypothesis by German (1968) suggested that the maternal age effect was attributable to the fact that coitus was less frequent with advancing maternal age and thus more likely to occur some time after ovulation. Presumably this delay in fertilization affected either the probability of second division non-disjunction in the fertilized ovum or survival of the resultant aneuploid zygote. The hypothesis appears unlikely to a account for the maternal age effect (see references in Smith and Berg, 1976, p.262) particularly because in at least 80% of cases the extra chromosome results from 1st division non-disjunction. But it cannot be dismissed as a contributing factor, independent of maternal age, to at least 2nd division maternal non-disjunction.

A related hypothesis was proposed (although never published to my knowledge) by Harlap. This was put forward as I recall to account for the high frequency of Down syndrome among the non-European Jews in the Israel study described above, the bulk of whom are very orthodox religiously. In this group there is often strict observance of the rules regulating the timing of coitus. These make it very likely that fertilization occurs around the time of ovulation. (Couples are not supposed to have coitus from the onset of menstruation until the wife undergoes a ritual bath which is to occur one week after the cessation of menstruation.) I tried to ascertain for investigation a population of orthodox Jews in this country who adhere to this practice, but for various reasons was not successful. I hope to pursue this lead elsewhere. Certainly maternal age specific rates in such populations similar to rates in others would be presumptive evidence against a factor involved in "delayed" or "late" fertilization of the ovum.

MATERNAL AGE EFFECTS

Maternal age accounts primarily and probably exclusively for the entire association of the 47,+21 genotype or the Down syndrome phenotype with parental age. Admittedly, there is some disagreement about a small paternal age effect under some circumstances. This has been commented upon extensively elsewhere in the literature (e.g. Stene et al, 1987 and Hook, 1987) and I won't consider it further here.

I will limit the bulk of discussion to changes of Down syndrome or 47,+21 with age because we have the most extensive data here. The rate undergoes an exponential increase above age 30. From 20 to 30 there is only a slight increase, with increasing age. Under age 20 there <u>may</u> even be a slight increase with <u>decreasing</u> age, so that at age 15 in a few studies the rate is roughly equal to that at age 30. (See references in Hook, 1981b or Hook, 1982.)

The younger age phenomenon may be simply the result of statistical fluctuation; it is not found in all data sets. Even if it is "real", it may have nothing to do with primary non-disjunction. Rather, in many populations teen-age pregnancies are relatively more prevalent among the retarded and the retarded are likely to include a higher proportion of trisomy 21 mosaics. Alternatively the phenomenon may result from random statistical errors. For example, random changes of a "2," "3" or "4" to a "1" in the first digit of the age (or related errors in birth year) would raise artifactually the rate in those believed to be under 20. Random changes in the other direction would have little counter-balancing effect because the rates are much higher at the older age. See Hook et al, 1984 for a detailed discussion of how such random effects can produce artefac-tual leveling effects.

At least five different statistical models have been proposed to account for the marked increased in rate above age 20 (Lamson and Hook, 1981; Hook et al, 1984):

1) A simple exponential increase with age superimposed upon a constant rate, i.e. $y = a + \exp(bx+c)$ where y = rate, x = maternal age and b and c are parameters. (b is about .25 to .30 in most data sets).

2) A Poisson increase with age plus a constant rate, i.e. $y = a$ + sum($i=n$ to infinity) [P(i,bx)], where P(i,bx) = [(bx) to the i power] multiplied by [e to the -bx power] divided by [i factorial], a, b and n are parameters of the equation and e is the base of the natural logarithm. (n varies from 17 to 25 in most date sets).

3) A "power" model superimposed upon a constant effect, i.e. $y = a + b$(x to the power n) where n is an integer. (n varies from 10 to 13 in most data sets.)

4) A divided slope model $\underline{y} = \exp[(\underline{b} + \underline{c})\,\underline{x} + \underline{a}]$ where $\underline{b}$ = zero above a "boundary" around 30 and $\underline{c}$ = zero below such a boundary. This model has not only the parameters $\underline{b}$, $\underline{c}$ and $\underline{a}$ but also an implicit parameter, the boundary at which the age transition occurs.

5) A more complex exponential regression model, e.g. $\underline{y} = \exp(\underline{d}(\underline{x}\text{ squared}) + \underline{cx} + \underline{b}) + \underline{a}$, or $\exp(\underline{f}(\underline{x}\text{ cubed}) + \underline{d}(\underline{x}\text{ squared}) + \underline{cx} + \underline{b}) + \underline{a}$, where $\underline{d}$ and $\underline{f}$ are also parameters.

The last two are rather uninteresting because they are associated with no obvious biological models and may have more than three parameters. I will not discuss them further although this does not mean they are necessarily "incorrect".

The first one is of great interest. An exponential increase is compatible with an infectious process. Penrose rejected this exponential model because he believed the plotted graph of rate against age levelled off (on an exponential scale) above age 40, at ages at which the rate of 47,+21 is no more than 1% or 2%. A true infectious process shouldn't level off until the rate reaches closer to 50% or so.

Penrose looked only at data by five-year intervals. Such analyses may have certain biases that falsely suggest such a leveling (Hook and Lamson, 1980). In data by one year interval I have not seen this levelling in U.S. data sets on live births or amniocentesis (Hook and Lamson, 1980; Hook et al, 1984). But the data of Ferguson-Smith and Yates (1984) from Europe do indicate a leveling off at age 47 and above, at least in amniocentesis data. More perplexing, the data on other trisomies, 47,+13 and 47,+18, in both U.S. and European amniocentesis data sets also suggest a leveling off, at ages 43 and 44, respectively, but only for these abnormalities. Perhaps this is explicable by higher earlier prenatal mortality at upper boundaries of age for 47,+13 and 47,+18. But let us put this aside for a moment. At least the data by one year interval on 47,+21 don't indicate as marked a leveling as Penrose inferred from the older five year date. Moreover, statistical fluctuations or random coding errors can result in leveling effects at the extremes of age even in one year analyses (Hook et al, 1984).

The power and Poissons model which Penrose preferred have slightly different etiological implications. A power model with threshold _n_ postulates say, some kind of single "structure" or process of unspecified type which fails after a succession of independent accidents associated with age and which, when greater or equal to some threshold _n_, causes meiotic division to become inaccurate. (Imagine a trunk of a tree which falls after _n_ blows of an axe.) Penrose visualized damage with age to the spindle mechanism as a plausible biological mechanism for such a model. On this view all damage is cumulative.

The Poisson model with parametic _n_ postulates that _n_ or more "deleterious events" of chance occurrence which affect say, one or the other centromere, cause non-disjunction. In particular, Penrose proposed breaks in strands attached to the centromere. (Visualize a scissors which must cut _n_ strands to have an effect.) The essential differences in the power and Poisson models as I understand them is that in the Poisson model damage is only cumulative if different strands are cut; cutting the same strand twice won't affect the process any more than cutting it once. Thus all "damage" is not necessarily cumulative. In the power model, consecutive blows of damage to the same structure have a cumulative effect up to the threshold.

Penrose rejected the exponential model because he got better fits to his data set with both the power model (for _n_=10) and with a Poisson model for (n=17). The Poisson model gave the best fit and he preferred the etiological hypothesis associated with it.

His discussion appears in his book with Smith (see p.167-170 in Penrose and Smith, 1966) and is carried over verbatim to the 2nd edition (Smith and Berg, 1976). He did not to my knowledge publish this anywhere else.

Lamson and Hook (1980 and 1981) applied the three different models proposed by Penrose (and one other) to five other data sets on 47,+21. If we consider only the three models discussed above, one data set fit the power model best (n=11), three fit the exponential model best, and one fit none of the models well but did least poorly with the Poisson model (n=25). (See Lamson and Hook, 1980 and 1981 for further discussion.)

It should be noted that increasing the value of n in either the power model, from n=10 or in the Poisson model, from n=17, will improve the fit to data sets which do not level off with age or level off less than the original data set of Penrose.

Lamson and Hook (1981) concluded that there were not enough data to choose either of these three models. A preference could only be based upon heuristic considerations. Adding more parameters will of course improve the fits of all models, but adds nothing to an etiological understanding.

One attraction of the exponential model is that it it implies a biological factor which may be specified in greater detail. An infectious model applied to non-disjunction would imply that the responsible agents spreads from oocyte to oocyte with age, infecting cells at a rate proportional to the number already affected up to some limit of saturation.

How would such an agent enter the ovary?

Obviously external infection of some type can always be invoked. But just as readily a retrovirus already in the genome could be postulated. Some factor may release or trigger the first activation of a retrovirus after which transfection of other cells may result in autocatalysis of release of additional latent viruses at an exponentially increasing rate. This process would be superimposed upon any age independent processes which are not dependent upon the viral mechanism.

Obviously, this is a very general hypothesis. But I find it very interesting that statistical considerations suggest a particular etiology -- a latent virus whose consequences increase with age -- which is subject to investigation. The hypothesis also has a least some implications for primary prevention of Down syndrome.

Perhaps about as much as we can ask of statistics is that it enables us to describe empiric risks and to test quantitative predictions of hypotheses. That it may also lead to a theory about a specific type of etiological mechanism is an unexpected bonus. But the final evaluation of this viral infection hypothesis can of course only come from direct biological investigation.

Lastly, the discussion above assumed that the bulk of the maternal age effect is attributable to effects of ageing upon the fidelity of chromosome disjunction in the female. Yet ageing effects could result less directly from selective effects at pre-gametic, gametic or post-gametic stages.

Pre-gametic effects could result from selective factors in the ovary that postpone the ovulation of aneuploid ova until advanced age, the so-called "production line" hypothesis of Henderson and Edwards; gametic effects could result from selective fertilization with advancing age of disomic gametes; and post-gametic effects could be due to preferential

TABLE 1. Estimates of cytogenetic abnormality associated with various types of morbidity and mortality. Estimates will vary with livebirth prevalence which is a function in part of the proportion of children born to older mothers and use of prenatal diagnosis and selective abortion. These estimates (particularly those on postnatal mortality and morbidity) are on populations around 1975 in developed countries. The proportion of mothers 35 yr or older was roughly 6-10%, with some variation among populations and temporal periods. Prenatal diagnosis and selective abortion was not widely used until later.

Conceptuses or individuals with:	Proportion with cytogenetic abnormality:
Early embryonic death in unrecognized pregnancies	Unknown, perhaps 33-67%
Recognized embryonic and fetal deaths (≥5 weeks of gestation)	About 30% total. Rate varies from 50% at 8 to 11 weeks to about 5% in stillbirths (≥28 weeks)
Infant and childhood deaths	5-7%
Structural congenital malformations	4-8%
Congenital heart defects	13%
Three independent birth defects and mental retardation retardation (all of unknown etiology)	5.5% (case rate minus control rate)

TABLE 1. (cont'd.)

Condition	Group	Frequency
Mental retardation (excluding fragile-X)	IQ < 20	? 3-10%
	IQ 20-49	12-35%
	IQ 50-69	? 3%
Other developmental disability		? 1-3%
Criminality (defined operationally by presence in security setting)	Males in "ordinary" prisons	0.8%
	Males in settings for criminal psychopaths, retarded criminals, etc.	3.0%
	Females in prison	0.4%
Infertility in males		2% (15% in those azoospermia)
Male pseudohermaphrodites		<25%
True hermaphrodism		25%
Multiple miscarriages in couples		2-5% (depending upon screening out of other causes)
Females with pubertal anomalies		27%
Females with hypergonadotrophic hypogonadism		65%

Adapted from Hook (1985a)

survival of the conceptus in an older woman -- the putative "relaxed selection" effect (Hook, 1985b). The evidence for these is discussed further by Warburton elsewhere in this volume. The proposed gametic and post-gametic mechanisms at least are inconsistent with the viral hypothesis suggested above.

REFERENCES

Alfi OS, Chang R, Azen SP (1980). Evidence for genetic control of non-disjunction in man. Amer J Hum Genet 32:477-483.

Angell RR, Templeton AA, Aitken RJ (1986). Chromosome studies in human in vitro fertilization. Hum Genet 72:333-339.

Awa AA, Honda T, Neriishi S, Sufuni T, Shimba H, Ohtaki K, Nakano M, Kodama Y, Itoh M, Hamilton HB (1987). Cytogenetic study of the offspring of atomic bomb survivers, Hiroshima and Nagasaki. In: Obe G, Busler A, eds. Cytogenetics: basic and applied aspects. Berlin, Springer-Verlag 166-183.

Baker BS, Carpenter ATC, Esposito MS, Esposito RE, Sandler L (1976). The genetic control of meiosis. Ann Rev Genet 53-134.

Brandriff B, Gordon L, Ashworth LK, Watchmaker G, Carrano AV (1986). Detection of chromosomal abnormalities in human sperm. Progr Clin Biol Res 209B:469-476.

Cuckle H, Wald N, Stone R, Densem J, Haddow J, Knight G (1988). Maternal serum thyroid antibodies in early pregnancy and fetal Down's syndrome. Pren Diag 8:439-445.

Daniel A, Hook EB, Wulf A (1988). Collaborative U.S.A. data on prenatal diagnosis for parental carriers of chromosome rearrangements: risks of imbalanced progeny. In: The cytogenetics of mammalian autosomal rearrangements, ed. A. Daniel. New York, Alan R. Liss, Inc. 73-162.

Ferencz C (1985). The etiology of congenital cardiovascular malformations: observations on genetic risks with implications for further birth defects research. J Med Clin Exp Theor 16:497-508.

Ferencz C, Rubin SD, McCarter RJ, Boughman JA, Wilson PD, Brenner II, Neill CA, Perry LW, Hepner SI , Downing JW (1987). Cardiac and non-cardiac malformations: observations in a population based study. Teratology 367-378.

Ferguson-Smith MA, Yates JRW (1984). Maternal age specific rates for chromosome aberrations and factors influencing them: report of a collaborative European study on 52,965 amniocenteses. Prenat Diag 4:5-44.

Forssman H, Akesson HO (1967). Consanguineous marriages and mongolism. In: Mongolism: Ciba Foundation Study Group No. 25., eds. Wolstenholme GEW, Porter R, London, J&A Churchill, 23-29. (See also the discussion following on pages 29-34.)

German J (1968). Mongolism, delayed fertilization, and human sexual behavior. Nature 217:516-518.

Grell RF (1962). A new model for secondary nondisjunction: The role of distributive pairing. Genetics 47:1734-54.

Grell RF (1971). Distributive Pairing in man? Ann Genet 14:165-171.

Grell RF (1985). The meiotic process and aneuploidy. In: Aneuploidy: etiology and mechanisms, eds. Dellarco VL, Voytek PE, Hollaender A, New York, Plenum 317-336.

Hecht F. (1977). The non-randomness of human chromosome abnormalities. In: Hook, EB, Porter, IH, eds. Population cytogenetics-studies in humans. New York, Academic Press, 237-250.

Heinonen OP (1976). Risk factors for congenital heart disease: a prospective study. In: Kelly, S, Hook, EB, Janerich, DT and Porter, IH, eds. Birth defects: risks and consequences. New York, Academic Press.

Hook EB (1974). Racial differentials in the prevalence rates of males with sex chromosome abnormalities (XXY, XYY) in security settings in the U.S.A. Amer J Hum Genet 26:504-511.

Hook EB and Lamson SH (1980). Rates of Down's syndrome at the upper extreme of maternal age - Absence of a leveling effect and evidence for artifacts resulting from an analysis of rates by five year maternal age intervals. Amer J Epid 111:75-80.

Hook EB (1981a). Prevalence of chromosome abnormalities during human gestation and implications for studies of environmental mutagens. Lancet 2:169-172.

Hook EB (1981b). Down's syndrome: Its frequency in human populations and some factors pertinent to variation in rates, in de la Cruz, FF and Gerald, PS eds., Trisomy 21 (Down Syndrome): Research perspectives, University Park Press, Baltimore, 3-67.

Hook EB (1982). The epidemiology of Down syndrome. In: Pueschel, SM, ed. Down syndrome: Advances in Biomedicine and the Behavioral Sciences. Cambridge, Ware Press 11-88.

Hook EB (1983). Perspectives in mutation epidemiology: 3. Contribution of chromosome abnormalities to human morbidity and mortality and some comments upon surveillance of chromosome mutation rates. Mutat Res 114:389-423.

Hook EB (1984). [The epidemiology of] human chromosome abnormalities. In: Bracken, MB ed. Perinatal Epidemiology, New York, Oxford University Press 3-22.

Hook EB, Cross PK, Regal RR (1984). The frequency of 47,+21, 47,+18 and 47,+13 at the uppermost extremes of maternal ages: Results on 56,094 fetuses studied prenatally and comparisons with data on livebirths. Hum Genet 68:211-220.

Hook EB (1985a). The impact of aneuploidy upon public health: mortality and morbidity associated with human chromosome abnormalities. In: Dellarco, VL, Voytek, PE, Hollaender, A, eds. Aneuploidy: etiology and mechanisms, New York, Plenum Press 7-33.

Hook EB (1985b). Maternal age, paternal age, and human aneuploidy: nature, magnitude, etiology and mechanism of effects. In: Dellarco, VL, Voytek, PE and Hollaender, A, eds. Aneuploidy: etiology and mechanisms. Plenum Press, New York 117-132.

Hook EB (1987). Issues in analysis of data on paternal age and 47,+21: Implications for genetic counseling for Down syndrome. Hum Genet 77:303-306.

Hook EB and Cross PK (1987). Rates of mutant and inherited structural cytogenetic abnormalities detected at amniocenteses: Results on about 63,000 fetuses. Ann Hum Gen 51:27-55.

Jacobs PA (1975). The load due to chromosome abnormalities in man: The possible evolutionary significance of chromosome abnormalities. In: Salzano FM, ed. The role of natural selection in human evolution. Amsterdam, North-Holland Publishing Co. pp.347-351.

Jacobs PA, Hunt PA, Matsuura JJ, Wilson CC, Szeulman AE (1982). Complete and partial hydatidiform mole in Hawaii: cytogenetics, morphology and epidemiology. Br J Ob Gyn 261-282.

Lamson SH, Hook EB (1980). A simple function for maternal age specific rates of Down's syndrome in the 20-49 age interval and its biological implications. Amer J Hum Gen 32:743-753.

Lamson SH, Hook EB (1981). Comparison of mathematical models for the maternal age dependence of Down's syndrome rates. Hum Genet 59:232-234.

Martin RH (1985). Chromosomal abnormalities in human sperm in Aneuploidy: etiology and mechanisms. In: Dellarco VL, Voytek PE, Hollaender A, eds. New York Plenum 91-102.

Matsunaga E. (1966). Down's syndrome and maternal inbreeding. Acta Genet Med Gemell 15:224-229.

Penrose LS, Smith GF (1966). Down's anomaly. London, Churchill pp.162-171.

Schmiady H, Kentenich H, Stauber M (1987). Chromosome studies of human in vitro fertilization (IVF) failures. In: Obe G, Basler A, eds. Cytogenetics:basic and applied aspects. Berlin, Springer-Verlag 184-197.

Sever JL, Gilkeson MR, Chen TC, Ley AC, Edmonds D (1970). Epidemiology of mongolism in the collaborative project. Ann NY Acad Sci 171:328-40.

Smith GF, Berg JM (1976). Down's anomaly. Edinburgh, Churchill-Livingstone.

Stene E, Stene J, Stengel-Rutkowski S (1987). A reanalysis of the New York State prenatal diagnosis data on Down's syndrome and paternal age effects. Hum Genet 77:299-302.

Stark CR, White NB (1977). Cluster analysis and racial differences in risk of Down's syndrome in Population Cytogenetics: studies in humans, eds. Hook EB, Porter IH, New York, Academic Press 275-84.

Sutherland GA (1900). The differential diagnosis of Mongolism and Cretinism. Lancet 1:23-24 (cited by Volpe 1986, op. cit.)

Volpe EP (1986). Is Down syndrome a modern disease? Persp Biol Med 29:423-436.

Warburton D (1987). Reproductive loss: how much is preventable? New Eng J Med 316:155-157.

Warburton D, Kline J, Stern Z, Hutzler M, Chin A, Hassold T (1987). Does the karyotype of a spontaneous abortion predict the karyotype of a subsequent abortion? - Evidence from 273 women with two karyotyped spontaneous abortions. Amer J Hum Genet 41:465-483.

Wramsby H, Fredga K, Liedholm P (1987). Chromosome analysis of human oocytes recovered from preovulatory follicles in stimulated cycles. New Eng J Med 316:121-124.

Zenzes MT (1987). Chromosome analyses using human spermatozoa. In: Obe G and Basler A, eds. Cytogenetics: basic and applied aspects. Berlin, Spring-Verlag 198-224.

Molecular and Cytogenetic Studies of
Non-Disjunction, pages 29–43

Molecular Mapping of Chromosome 21 and the region responsible for Down Syndrome

Stylianos E. Antonarakis[1], Andrew C. Warren[1], Mary Kay McCormick[1], John G. Lewis[1], Philip A. Hieter[2], Aravinda Chakravarti[3]

[1]Genetics Unit, Department of Pediatrics, The Johns Hopkins University School of Medicine, Baltimore, MD 21205, [2]Department of Molecular Biology and Genetics, The Johns Hopkins University School of Medicine, Baltimore, MD 21205, [3]Department of Biostatistics, University of Pittsburgh, PA.

INTRODUCTION

Chromosome 21 is the smallest human chromosome comprising about 1.4% of the total human genomic length. Trisomy 21 (Down Syndrome) is the major abnormality associated with this chromosome and is the most commonly known genetic cause of mental retardation (Dela Cruz & Gerald eds. 1981). A large number of single copy DNA sequences have been assigned to chromosome 21 (Human Gene Mapping 9). Using several somatic cell hybrid lines these DNA sequences have been regionally mapped to more than 10 subsegments of this chromosome (see Van Keuren et al 1986 for example). Irradiation of hybrid cell lines also contributed to further mapping of the long arm of the chromosome (Cox et al 1988). Pulsed field gel electrophoresis allowed further subgrouping of chromosome 21 probes in terms of large macrorestriction fragments (Abad et al 1988). Using the CEPH families and a large Venezuelan pedigree initial linkage maps for chromosome 21 have been constructed (Tanzi et al 1988, Donis-Keller et al 1987). The development of cloning of large DNA fragments (100+ kb) in yeast artificial chromosomes (YACs) (Burke et al 1987) will aid in the construction of physical maps of human chromosomes including chromosome 21. In this paper a linkage map of 17 DNA loci on chromosome 21 will be described. Initial attempts (i) to define the minimum chromosomal region which if triplicated is responsible for Down Syndrome and (ii) to clone chromosome 21 in YACs will also be presented.

A. GENETIC LINKAGE MAP OF HUMAN CHROMOSOME 21

A1. Materials & Methods

Pedigrees: Human DNA samples from two or three generation families, typically comprising grandparents, parents and at least six children were provided by CEPH (Centers d'Etude du Polymorphism Humaine, Paris) (Dausset 1986). All 40 families were used in our analyses. Genotypic data for five loci (see table 1) were studied in 23 CEPH families by Donis-Keller et at 1987 and their data were made available to us for combined analysis.

Restriction analysis and DNA probes: Genomic DNA was digested with various restriction enzymes and the DNA fragments were transferred to nitrocellulose filters and hybridized to different DNA probes as described. The loci included in the linkage analysis are shown in Table 1.

TABLE 1

GENETIC MARKERS STUDIED ON HUMAN CHROMOSOME 21 IN THE CEPH PEDIGREES

LOCUS NAME	POLYMORPHISM	SOURCE/REFERENCE
D21S13	Taq I	KE Davies (Hum Genet 66:54, 1984)
D21S1	Bam HI	PC Watkins (NAR 13:6075, 1985)
D21S11	EcoRI	PC Watkins (NAR 13:6075, 1985)
D21S3	Hind III, TaqI	PC Watkins (NAR 13:6075, 1985)
D21S111	Sst I	C Wong (Proc Natl Acad Sci USA, in press, 1989)
D21S8	Hind III	PC Watkins (NAR 13:6075, 1985)

* BCEI	Bam HI	P Chambon (NAR 10:7895, 1982)
* D21S112	Rsa I VNTR	H Donis-Keller (Cell 51:319, 1987)
* D21S15	Msp I	G Stewart (Cytog Cell Genet. 46:257, 1987)
* D21S17	Bgl II	G Stewart (Cytog Cell Genet. 46:257, 1987)
* SOD1	Msp I	Y Groner (EMBO J 4:77, 1985)
ETS2	Msp I	T Papas (Proc Natl Acad Sci USA 82:7294, 1985)
COL6A1	TaqI VNTR	ML Chu (Am J Hum Genet 42:435, 1988)
APP	Not I/Bgl II	N Robakis (Genomics 1:307, 1987)
D21S82	EcoRI VNTR	G Scherer (NAR 15:5499, 1987)
PFKL	Kpn I	Y Groner (BBRC 141:374, 1986)
D21S110	Msp I	D Kurnit (Cytog Cell Genet 46:257, 1987)

* Genotypic data for these loci were obtained by Collabrative Research, Inc. (Cell 51:319, 1987)

Linkage analysis: Two point and multipoint analysis under the assumption of no interference (Haldane map function; Haldane 1919) was performed using the computer programs LINKAGE version 4.7 (Lathrop & Lalouel, 1988) and MAPMAKER version 1.0 (Lander et al 1987)

A2. Results

Next figure shows the genetic map of human chromosome 21 for male and female meioses. The markers have been placed on the map distance scale assuming no interference together with the recombination values for each segment. The physical locations of selected markers are also shown and demonstrate the correspondence between the physical and the genetic maps.

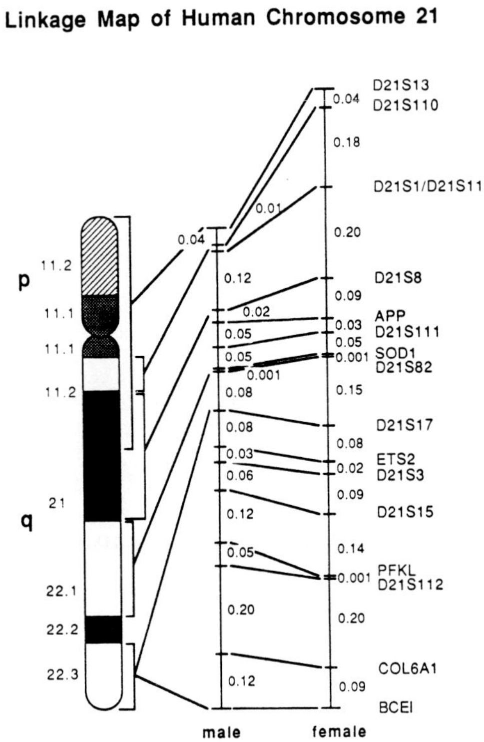

Figure 1: Primary genetic map for human chromosome 21 in males and females with physical locations of selected markers; the scale is in CentiMorgans. Recombination values for adjacent segments is each sex are indicated.

Table 2 presents the total calculated length of the map in males and females and the sex-average map that assumes no difference in recombination between sexes. Lengths for the two former maps were calculated under the assumptions of a constant ratio of male:female map distance and for a ratio that varied between different locus pairs. Calculations under the assumption of chiasma interference according to the Kosambi (1944) model and for no interference (Haldane 1919) are also shown.

TABLE 2

MAP LENGTHS IN CM CALCULATED UNDER DIFFERENT ASSUMPTIONS OF SEX SPECIFIC DIFFERENCES IN MAP DISTANCE AND TWO MAPPING FUNCTIONS.

	MAP FUNCTION	
	Haldane	Kosambi
Sex Average Map	136 ± 45	121 ± 41
Constant Ratio: Male Map	110 ± 52	100 ± 47
Female Map	167 ± 57	146 ± 52
Variable Ratio: Male Map	114 ± 53	103 ± 48
Female Map	160 ± 55	140 ± 50

A3. Discussion

The genetic data presented, together with the physical localizations of markers, produces a map over 100 cM long that covers almost the entire length of the long arm of human chromosome 21. This map is approximately twice the genetic length calculated from estimates of the physical length of the chromosome, and assuming constant recombination rates along the chromosome (Laurie and Hulten 1985). Our data suggest, however, that recombination rates do vary along the chromosome and between sexes, with females having approximately twice as much recombination as males between proximal markers, although recombination rates are similar between distal markers between sexes (Figure 1). Our data also indicate that about half the crossovers occur distal to ETS2 which has been physically mapped to the terminal band 21q22.3, in agreement with the findings of Tanzi et al (1988). This is also in agreement with physical data indicating that chiasmata are most commonly found on the distal part of chromosomes 21 (Laurie and Hulten 1985). Thus, over 50% of the current genetic map of chromosome 21 is localized to only about 15% of the physical length of this chromosome (band 21q22.3). The genetic length of this terminal band may be even greater, and can only be estimated once telomeric sequences are mapped.

B. DEFINITION OF THE MINIMUM CHROMOSOMAL REGION RESPONSIBLE FOR THE DOWN SYNDROME PHENOTYPE (DS-21MCR)

Cytogenetic analyses of partial trisomy 21 patients demonstrated that the presence of three copies of only chromosomal band 21q22 was necessary to result in the phenotype of DS (Dela Cruz & Gerald 1981) suggesting that the gene products of specific normal genes in this area were contributing to this phenotype. More precise cytogenetic localization of the "Down Syndrome" region of chromosome 21 was controversial due to the description of patients that either included or excluded subbands 21q22.1 and 21q22.3 (Habedank et al 1982, Poissonnier et al 1976). Therefore we have undertaken the molecular characterization of the DS region of chromosome 21 with the intention of defining a minimal chromosome region (MCR) that results in a DS phenotype when present in three copies in an individual. This approach was facilitated by the availability of chromosome 21 specific single copy DNA probes and both physical and genetic maps of the chromosome.

B1. Methods

DNA probes and determination of copy number of loci: Sequences from the following loci have been used as probes to determine copy numbers of chromosome 21 in the patients: D21S13, D21S1, D21S11, D21S8, APP, D21S111, SOD1, D21S58, D21S82, D21S55, D21S57, D21S3, ETS2, PFKL, D21S42, D21S39, D21S112, D21S56, COL6A1, COL6A2. DNA polymorphisms have been reported for most of these probes (Human Gene Map 9). Control probes used were sequences from the ß-globin locus on chromosome 11p or the APOA1 locus on chromosome 11q (Human Gene Map 9). The copy number of each probe is determined in each patient by DNA polymorphism analysis. Genomic DNA was analyzed by southern blot as described (Southern 1975). Copy number of the sequence was determined by densitometric comparison of the intensity of the allelic bands on autoradiographs. If an individual was not informative for a polymorphism, then copy number was determined by dosage analysis (comparison of intensity of alleles relative to that of a control non-chromosome 21 probe by densitometry.)

Patients: Sixteen patients were characterized in this manner. Table 3 lists the clinical diagnosis of the patients and their karyotype. (Our minimal criteria for the phenotype of Down Syndrome are mental retardation and classic facial features.)

TABLE 3
PATIENTS EXAMINED FOR THE DS-21MCR*

	Patient	Clinical Diagnosis	Cytogenetic Description
1	MP02	"DS" phenotype	normal karyotype
2	DD01	"DS" phenotype	normal karyotype
3	DD02	"DS" phenotype	normal karyotype
4	AS06	"DS" phenotype	normal karyotype
5	AS04	DS phenotype	dup 21q22
6	AS01	DS phenotype	dup 21q22.2-qter (Hum.Genet. 60:74, 1982)
7	AS02	DS phenotype	dup 2q22.2-qter (Hum.Genet. 60:74, 1982)
8	AS03	DS phenotype	dup 21q22.2qter (Hum.Genet. 60:74, 1982)
9	GM3997	DS phenotype	dup 21q21-22? (Clin.Genet. 24:97,1983)
10	GM4108	DS phenotype	dup 21q11-qter (J.Med.Genet. 19:144, 1982)
11	AS05	DS phenotype	dup 21q22 (Arch.Genet. 52:171, 1979)
12	GM1413	DS phenotype	dup 21q21.05-qter (Am.J.Hum.Genet.27:478, 1975)
13	MP01	DS phenotype	dup 21q22
14	EC01	no DS phenotype	dup 21q22
15	GM5881	no DS phenotype	+21q-?
16	MP03	no DS phenotype	dup 21q11.2-q21.2

Samples from patients 1, 13 and 16 have been supplied by Dr. M. Mikkelsen, L. Tranebjaerg, M. Petersen of the John F. Kennedy Institute, Department of Medical Genetics, Glostrup, Denmark.

Samples from patients 2 and 3 have been supplied by Dr. D. Driscoll of the Johns Hopkins University School of Medicine, Baltimore, MD 21205

Samples from patients 4, 5, 6, 7, 8 and 11 have been supplied by Dr. A. Schinzel, Institute for Medical Genetics, University of Zurich, Switzerland and Dr. M. Habedank, Aachen, FR Germany

Sample from patient 14 has been supplied by Dr. E. Cantu, Department of Pediatrics, Genetics Division, University of Florida, Gainesville, FL.

Samples from patients 9, 10, 12, and 15 have been supplied by the NIGMS repository.

The patients karyotypes have been examined in the laboratories of the investigators who supplied DNA or cell cultures.

B2. Results

The dosage of loci on chromosome 21 was determined in the patients using 18-20 single copy polymorphic DNA sequences as probes. The probes detect loci spread over the whole length of the long arm of chromosome 21. The results of polymorphism and dosage analysis for each of these probes in each of the patients is shown in figure 2 and are fully described in McCormick et al 1989. There is no molecular evidence for partial trisomy 21 in ten of sixteen patients (1-10), yet six of these ten (5-10) have cytogenetically visible chromosomal abnormalities described as partial trisomy 21. There is molecular evidence for partial trisomy 21 in six patients (11-16). Three of these have partial trisomy 21 for the distal long arm (11-13). The smallest triplicated area in common among them is distal to and including locus D21S55. The other three patients (14-16) have partial trisomy 21 for the proximal long arm, which can extend as far distal as locus D21S82 without resulting in a DS phenotype. Of the six patients with molecular partial trisomy 21, the area of transition between two and three copies in three patients occurs between loci D21S55 and D21S82. The genetic distance between these loci is approximately 12cM (Warren et al) and the physical distance is estimated to be approximately 10Mb.

In four patients (1-4) described as having a DS phenotype and normal karyotype, we have found no molecular evidence for partial trisomy 21. In addition, six patients

(5-10) are described as having a DS phenotype and partial trisomy 21, yet we have found no molecular evidence for partial trisomy 21 in any of these six patients indicating that the cytogenetic data and the molecular data are discrepant.

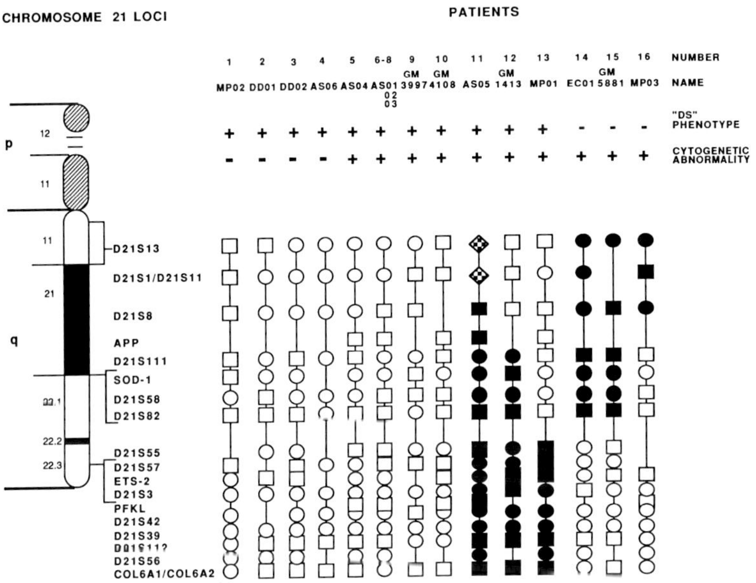

Figure 2. Molecular analysis of copy number of chromosomal loci in patients. White symbols indicate two copies of a sequence. Black symbols indicate three copies of a sequence. Squares indicate the information was obtained by DNA polymorphism analysis. Circles indicate the information was obtained by dosage analysis. Diamonds indicate only one copy of a sequence. Patients are designated by a name and a number. References to patients in the text will be by number for clarity. The presence (+) or absence (-) of a DS phenotype and a cytogenetically described abnormality is indicated under the patient name. The loci from which probes were derived are listed next to chromosome 21. The order shown was determined from the genetic linkage map for the long arm of chromosome 21 (Warren et al) except for loci D21S42 and D21S39. Their physical location in chromosomal band 21q22.3 was determined from somatic cell hybrids (unpublished observations), but their linkage relationship to the other

loci is unknown.

B3. Discussion

Analysis of DNA sequence copy number has been applied to the characterization of patients with partial trisomy 21 in order to molecularly define the "DS region" of chromosome 21. The most surprising result was the discrepancy between the cytogenetic data and the results of DNA hybridization analysis in patients 5-10. In these patients, the cytogenetic results demonstrated a visible chromosomal abnormality, presumed to involve chromosome 21 because of the phenotype of the patient. However the molecular evidence indicated two copies for all sequences examined without any suggestion of partial trisomy 21. A possible explanation for this is that the observed cytogenetic abnormality involves a segment of chromosome 21 for which there are no probes available, such as the region between loci D21S55 and D21S82. This is unlikely in some of the patients, such as 5-7, where the described abnormality is much larger than the area between D21S55 and D21S82. Alternatively, the described cytogenetic abnormality does not involve chromosome 21, and these patients are phenocopies of DS. There is also no molecular evidence for partial trisomy 21 in patients 1-4, who have normal karyotypes and several features compatible with a DS phenotype. It is possible that some or all of these four patients have partial trisomy 21 in an area which is not detected by probes currently available. In the event that no molecular evidence for partial trisomy 21 is obtained for any of these patients, then they are also phenocopies of DS and their phenotype can be ascribed to other genetic or environmental influences. The results in patients 11-16 allow a molecular definition of the DS region. The data in patients 14-16 suggest the area between loci D21S13 and D21S82 can be excluded from significant contribution to a DS phenotype since these patients have three copies for some or all of this region while having few if any features compatible with DS. The data in patients 11-13 might suggest that the DS-21 MCR is the smallest area present in common in three copies in these patients; that is, between and including loci D21S55 and COL6A1. This definition of the DS-21 MCR may need to be modified after the determination of the contribution of the regions between loci D21S55 and D21S82 and between COL6A1 and 21qter to a DS phenotype. The results in patients 13-15 suggest that the region between loci D21S55 and D21S82 may be important in contributing to a DS phenotype when present in three copies. If sequences from this region were found to be present in three copies in some of the patients, then the DS-21 MCR would need to be modified accordingly. The current DS-21 MCR necessary to result in a DS phenotype may also be affected by the characterization of a DS patient carrying an unbalanced t21;21 translocation fused at the 21q telomeres. This information would be important in determining the necessity of the most distal loci to a DS phenotype when present in three copies in an individual. If some of the most telomeric sequences are present in less than three copies in this patient, then the DS-21 MCR would not extend as far distal as it does currently. A final definition of the DS-21 MCR will require a more uniform distribution of DNA probes to minimize gaps in the molecular characterization, and a larger sample size to support the validity of a DS-21 MCR in many patients with partial trisomy 21 and a DS phenotype. Patients similar to our patient 13 have recently been reported. (Korenberg et al 1988, Sacchi et al 1988). The generation of a chromosome 21 specific library in yeast artificial chromosomes will provide the material for producing a uniform probe distribution as well as a source for ultimately identifying genes in the MCR which may be important in contributing to a DS phenotype.

C. CONSTRUCTION OF A PHYSICAL MAP OF HUMAN CHROMOSOME 21 IN YEAST ARTIFICIAL CHROMOSOMES

Yeast artificial chromosomes (YACs) are linear molecules composed of exogenous DNA and the fundamental functional components of a yeast chromosome which allow the artificial chromosome to be maintained in the host. Small linear chromosomes (<50 kb) are mitotically unstable, so that YACs must be ≥50 kb to be stably maintained. The feasibility of constructing YACs containing human DNA and the ability to generate YACs 500 kb in length was shown by Burke et al, 1987. This made it possible to consider the construction of chromosome specific libraries in YACs for the purpose of generating physical maps of chromosomes. The combined information from physical maps and genetic maps is critical to the ability to walk from clones to disease loci. YAC libraries of human chromosomes, by offering a 10 fold increase in clone size, over previous technology will contribute significantly to the physical mapping of the human genome and ultimately to the identification of disease loci. In particular, a chromosome 21 library in YACs will provide the material for defining the still uncharacterized region of chromosome 21 that may be important in contributing to a DS phenotype when present in three copies in an individual.

C1. Methods

The source of DNA for construction of a chromosome 21 library in YACs can be a hybrid cell line containing chromosome 21 as the only human chromosome or flow sorted chromosome 21. High molecular weight DNA is prepared in agarose blocks, partially or completely restricted with endonucleases and ligated to a YAC vector. The vector contains a yeast centromere for proper segregation of the YAC during mitosis and meiosis, yeast telomeres to allow complete replication of a linear molecule, and two ARS sequences for efficient and autonomous replication of the YAC. In addition, the vectors contain 2 selectable auxotrophic markers, a tRNA suppressor gene which allows assay for mitotic stability of the YAC by color phenotype, ateteretal a polylinker region for flexibility in cloning, an E.coli origin of replication, ampicillin gene and unique restriction sites for recovery of the ends of the YAC as plasmids in E.coli and a T7 promoter adjacent to the polylinker allowing ends of the YAC to be labelled and used as probes (Shero et al 1988). Following ligation to the vector, products are size selected on a low melting temperature agarose OFAGE gel, melted and transformed into yeast spheroplasts. Transformants are selected on plates lacking uracil and tryptophan and screened for mitotic stability. Stable transformants are screened by colony hybridization and/or by Southern blot of pulsed field gels hybridized to human specific repetitive probes. The ends of human containing YACs, are recovered as plasmids in E.coli, labelled and shown to map on chromosome 21.

C2. Results

The efficiency of transformation with control plasmid or small linear molecules is $1x10^6$ cfu/ug and $1x10^2$ - $1x10^3$ cfu/ug for ligation products. Stable recombinants represent between 60 and 95% of transformants. Approximately 3% of transformants originating from a hybrid cell line contain human DNA. One of these chosen for further characterization has been mapped to chromosome 21. The average size of YACs generated from EcoRI partial digests of hybrid DNA is ~150Kb (three different such YACs are shown in figure 3) and from Not I total digests of flow sorted chromosomes is ~350Kb.

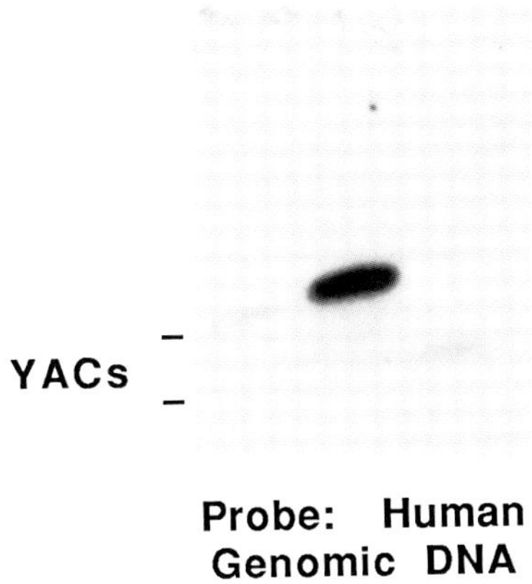

Figure 3. Autoradiogram of yeast artificial chromosomes after pulsed field gel electrophoresis. In each have a different YAC is depicted. The sizes are between 100 - 150 kb.

C3. Discussion

Chromosome 21 is ~4.5 x 10^4 kb. If the average size YAC is 150 kb, then ~1000 chromosome 21 YACs are needed to represent the chromosome three times. At current transformation efficiencies, this will require screening of 50,000 cfu using 50-100 ug of hybrid DNA as source material. Using flow sorted material as a source of DNA instead of genomic DNA from a hybrid has the advantage of reducing the number of primary transformants required to represent a library because 70% of these will have originated from chromosome 21 compared to 3% from genomic DNA from a hybrid cell line.

Constructing a physical map of chromosome 21 can be accomplished using a library generated from either source of DNA. Linking adjacent YACs can be done in 2 steps. The availability of somatic cell hybrid panels that divide chromosome 21 into several segments allows regional mapping by Southern blot of YACs prior to linking. YACs within a group can be ordered first, followed by linking of regional groups. The information generated from a physical map of chromosome 21 can be used, for example, to complete the molecular analysis of the "DS region" of chromosome 21.

ACKNOWLEDGMENTS

We thank Drs. D. Cox, H. Donis-Keller, P. Green, Y. Groner, L. Karathanasis, M.L. Chu, T. Papas, N. Robakis, G. Scherer, P. Watkins, J. Gusella, D. Kurnit, K. Davies, for providing DNA probes. We are greatful to Drs. A. Schinzel, M. Habedank, M. Petersen, G. Stetten, D. Driscoll, E. Cantu, L. Tranebjaerg, M. Mikkelsen for providing us with patient material and karyotypic analyses. We also thank S. Slaugenhaupt, M. Lathrop, J.M. Lalouel, E. Lander, K. Buetow for help in data analysis. Mr. J. Shero is especially acknowledged for providing us with plasmids for YAC cloning. The projects have been supported by NIH grants to SEA and PAH. We thank C. Wesko for expert secretarial assistance.

REFERENCES

Abad JP, Saito A, Ohki M, Pevny L, Cantor CR, Smith CL (1988) Program towards the completion of a restriction map of human chromosome 21. Am J Hum Genet. 43:A134 (abstract).

Botstein D et al (1987) A genetic linkage map of the human genome. Cell 51:319-337.

Burke DT, Carle GF, Olson MV (1987) Yeast artifical chromosomes with human DNA inserts. Science 236:806-811.

Cox DR, Price ER, Burmeister M, Sheffield V, Murray C, Ughim E, Myers RM (1988) Fine structure genetic analysis of human chromosome 21 using radiation hybrid mapping. Am J Hum Genet 43:A141 (abstract).

Dausset J. (1986) Le Centre d'Etude du Polymorphism Humaine - La Press Medicale 15:1801-1802.

Delacruz FF & Gerald PS (1981) Down Syndrome; Research perspectives. Baltimore: Univ Park Press.

Donis-Keller H, Green P, Helms C, Cartinhour S, Weiffenbach B, Stephens K, Keith TP, Bowden DW, Smith DR, Lander ES,

Habedank M & Rodewald A (1982) Moderate Down Syndrome in three siblings having partial trisomy 21q22.3 qter and therefore no SOD1 excess. Hum Genet 60:74-77.

Haldane J. (1919) The combination of linkage values, and the calculation of distances between the loci of linked factors. J Genet 8:299-309.

Human Gene Mapping 9 Paris Conference. (1987) Cytogenet Cell Genet 46:1-750.

Korenberg JR, Pulst SM, Kawashima H, Ikeuchi T, Yamamoto K, Ogasawara N, Schonberg SA, West R, Kojis T, Epstein CJ (1988) Familial Down Syndrome with normal karyotype; molecular definition of the region. Am J Hum Genet 43:A110 (abstract).

Kosambi D (1944) The estimation of map distances from recombination values. Ann Eugen 12:172-175.

Lander E, Green P, Abrahamson J, Barlow A, Daly M, Lincoln S, Newburg L (1987) MAPMAKER:An interactive computer package for constructing primary genetic linkage maps of experimental and natural populations. Genomics 1:174-181.

Lathrop GM. Lalouel JM (1988) Efficient computations in multilocus linkage analyses. Am J Hum Genet 42:498-505.

Laurie D, Hulten M, (1985) Further studies on bivalent chiasma frequency in human males with normal karyotypes. Ann Hum Genet 49:189-201.

McCormick MK, Schinzel A, Petersen MB, Stetten G, Driswll DS, Cantu ES, Tranebjaerlg L, Mikkelson M, Watkins PC, Antonarakis SE (1989) Molecular genetic approach to the characterization of the "Down Syndrome Region" of chromosome 21 Genomics, in press.

Poissonnier M, Saint-Paul D, Dutvillaux B, Gruyer P, Blignieve-Stronk G, (1976) Trisomic 21 partielle (21q21-21q22.2) Ann Genet 19:69-73.

Sacchi N, Nablantoglu J, Sergovich FR, Papas TS (1988) Human ETS2 gene on chromosome 21 is not rearranged in Alzheimer disease. Proc Natl Acad Sci USA 85:7675-7679.

Shero J, McCormick MK, Antonarakis SE, Hieter P (1988) Construction of vectors for yeast artificial chromosomes (submitted).

Southern EM (1975) Detection of specific sequences among DNA framents seqarated by gel electrophoresis. J Mol Biol 98:503-517.

Tanzi R, Haines J, Watkins PC, Stewart GD, Wallace MR, Hallewell R, Wong C, Wexler NS, Conneally PM, Gusella JF (1988) Genetic linkage map of human chromosome 21. Genomics 3:129-136.

Van Keuren ML, Watkins PC, Drabkin HA, Jabs EW, Gusella JF, Patterson D (1986) Regional localization of DNA sequences on chromosome 21 using somatic cell hybrids. Am J Hum Genet 38:793-804.

Warren A C, Slaugenhaupt S, Lewis JG, Chakravarti A, Antonarakis SE, (1988) A genetic linkage map of human chromosome 21 (submitted).

Molecular and Cytogenetic Studies of
Non-Disjunction, pages 45–79

GENE-CENTROMERE MAPPING AND THE STUDY OF NON-DISJUNCTION IN AUTOSOMAL TRISOMIES AND OVARIAN TERATOMAS

Aravinda Chakravarti, Partha P. Majumder, Susan A. Slaugenhaupt, Ranjan Deka, Andrew C. Warren, Urvashi Surti, Robert E. Ferrell, Stylianos E. Antonarakis

Human Genetics Division (A.C., P.P.M., S.A.S., R.D., R.E.F), Department of Pathology (U.S.), University of Pittsburgh, Pittsburgh, PA 15261, and, Department of Pediatrics (A.C.W.,S.E.A.), Johns Hopkins Hospital, Baltimore, MD 21205

INTRODUCTION

The genetic control of the different stages of meiosis is now well established and has been excellently reviewed by Baker et al. (1976). Studies of meiotic mutants in experimental organisms have led to three major generalizations. First, the enzymatic machinery necessary for meiotic development and recombination are not unique to germ cells but are also used by somatic cells in processes necessary to effect chromosome stability. Second, both the distribution and frequency of exchanges along the length of a chromosome are under specific and separate genetic control. Third, in chiasmate species, at least one exchange is necessary for normal chromosome disjunction, although not sufficient to insure normal meiotic segregation. These features emphasize the central role of crossing-over and genetic recombination in meiosis. Furthermore, recombination has great evolutionary importance because of its role in generating genetic variability and in the evolution of sexual reproduction (Maynard Smith, 1978).

The genetic complexity of meiosis suggests that abnormalities in various processes may lead to abnormal chromosome segregation, or non-disjunction. Among these, failure of chiasma formation (Darlington, 1929; Dobzhansky, 1933) or alterations in the distribution and frequency of chiasmata (Merriam and Frost, 1964) may be particularly important. In the mouse, meiotic non-disjunction is usually due to aberrant segregation of univalents and rarely due to failure of separation of homologs (Henderson and Edwards, 1968). These univalents may arise either by asynapsis (failure of chiasma formation during the zygotene-pachytene stage) or by desynapsis (premature unpairing of homologs following normal chiasma formation) (Henderson, 1970). However, in the Chinese hamster there is no correlation between univalent formation and subsequent non-disjunction at meiosis I (Sugawara and Mikamo, 1983).

In humans, it is not possible to study chiasmata on chromosomes that have undergone non-disjunction. However, genetic methods may be used to study the surrogate event of recombination, since chiasma formation and recombination are intimately involved (Polani and Crolla, 1982). Then, non-disjunction due to asynapsis or desynapsis is equivalent to no recombination or normal recombination, respectively. Alternatively, non-disjunction could result from an increased number of exchanges and subsequent chromosome entanglement; this may retard the separation of bivalents at anaphase I and lead to increased recombination (Bridges, 1916). The relative roles of these various processes may be estimated by studying the frequency of recombination between genetic markers on chromosomes that have undergone non-disjunction and by comparing them to control values (Chakravarti and Slaugenhaupt, 1987). In humans, the control values may be obtained from linkage studies in normal families (White et al., 1986) or from non-disjoined chromosomes arising from meiosis II errors which are expected to be unrelated to recombination. Alternatively, when studying trisomies, control values may also be obtained from the single normally disjoined chromosome in the trisomy (Halloran and Chakravarti, 1987).

This paper is an exposition of various aspects of gene-centromere mapping. First, we briefly review the types of non-disjunction products that may be studied in humans. Second, we outline the genetic and mathematical methods for gene-centromere linkage studies involving one or more genetic markers. Third, we discuss the types of genetic maps and genetic markers that are useful for these studies. Fourth, the statistical methods for gene-centromere mapping are presented. Finally, these methods are applied to data on ovarian teratomas and trisomy 21 in humans.

NON-DISJUNCTION IN HUMANS

Non-disjunction of chromosomes usually results in germ cells with one fewer chromosome or one extra chromosome compared to the normal complement. When these germ cells are fertilized, the zygote is either monosomic or trisomic. In humans, monosomy is virtually unknown, presumably being incompatible with early survival. However, trisomies are the most frequently identified chromosome abnormality, occurring in at least 4% of all clinically recognized pregnancies (Hassold and Jacobs, 1984). Although trisomies are usually associated with a single additional chromosome, double trisomies and mosaic trisomies have also been reported. Non-disjunction in human trisomies may occur either at meiosis I or at meiosis II, and in either the male or female parent; the most common errors occur at maternal meiosis I. The probabilities of the four modes of origin have been estimated for several autosomal and sex chromosome trisomies (Hassold and Jacobs, 1984; Hassold, Chiu and Yamane, 1984).

Non-disjunction may also involve the entire chromosome complement, in which case the germ cell is either empty, diploid, or rarely, polyploid.

Fertilization of an empty oocyte by a single sperm with subsequent endoreduplication of the paternal genome or fertilization with two sperm may lead to formation of a hydatidiform mole (Kajii and Ohama, 1977; Therman, 1986). Hydatidiform moles have a frequency of 1/2,000 pregnancies, and have received interest since they account for more than 50% of choriocarcinomas (Therman, 1986). An oocyte, when diploid, may be fertilized to produce triploidy, which is rare among livebirths (.01%) but comprises 17% of heteroploid spontaneous abortions (Carr and Gedeon, 1977). Alternatively, the unfertilized diploid oocyte may develop into an ovarian teratoma.

Human ovarian teratomas, also known as dermoid cysts, are the most common benign ovarian tumors and account for 11% of all ovarian tumors (Peterson et al., 1955). They develop from a totipotent germ cell and contain fully differentiated mature tissues, often from all three cell layers. The frequency of ovarian teratomas in the general population is unknown, but we have obtained a preliminary estimate of 0.2 - 0.6% by reviewing clinical records from 1972-1985 at Magee-Womens Hospital, Pittsburgh. The 630 cases identified during this period occurred in women with a median age of 30 years (range 15-80 years), of which 8.1% occurred bilaterally and 91.9% occurred unilaterally. The age distributions of unilateral and bilateral cases were not different.

Cytogenetic analysis of ovarian teratomas reveals a normal 46,XX karyotype, however, mosaicism, trisomy and structural rearrangements have also been observed (Ohama, 1987; Hoffner and Surti, 1989). Linder (1969) was the first to propose that ovarian teratomas are parthenogenic and arise by non-disjunction. Originally, all ovarian teratomas were thought to arise by meiosis II errors (Kaiser-McCaw et al., 1976); however, diverse origins of teratomas even in a single individual have since been demonstrated (Carritt et al., 1982; Parrington et al., 1984). Surti et al. (1988) have postulated five different mechanisms of origin: (1) failure of meiosis I or fusion of the first polar body with the oocyte; (2) failure of meiosis II or fusion of the second polar body with the oocyte; (3) endoreduplication of the genome of a mature ovum; (4) failure of meiosis I and II in a primordial germ cell; and (5) fusion of two ova.

The above brief review emphasizes the varied types of non-disjunction products that may be observed in humans. Of these, trisomies, triploids and ovarian teratomas arising by failure of meiosis I or II are useful for studying the effect of recombination on non-disjunction.

GENE-CENTROMERE MAPPING

Background

Genes may be mapped relative to the centromere whenever two or more members of a meiotic tetrad can be recovered. In humans, non-disjunction usually results in an unordered half-tetrad containing two of the four chromatids. The non-disjunction event can occur at meiosis I or II and can be inferred from the centromeric genotype. In the absence of recombination, a meiosis I error leads to a disomic gamete that is heterozygous at the centromere and at all genetic markers along a chromosome arm, since non-sister chromatids are involved; under meiosis II non-disjunction, the centromere and genetic markers are homozygous since sister chromatids are involved. These disomic gametes are also called bi-grandparental and uni-grandparental, respectively, since non-sister chromatids (derived from two grandparents) and sister chromatids (derived from a single grandparent) are involved (Côté and Edwards, 1975). With the occurrence of recombination between genetic markers and between the markers and the centromere, meiosis I and II errors may produce homozygous and heterozygous genetic markers, respectively, the probability of these events increasing with increasing genetic distance between the centromere and the marker loci. Under recombination, then, a disomic gamete may be uni- or bi-grandparental in different segments of the same chromosome.

Probability of Nonreduction

Mapping genes relative to the centromere requires determination of the numbers of non-disjunction products which are heterozygous and homozygous at any marker locus which is heterozygous in the host and whether they arose as meiosis I or II errors. The probability of retained heterozygosity at a marker locus is called the probability of nonreduction. If c crossovers occur between the centromere and a marker locus, the probability of nonreduction for meiosis I non-disjunction (MI NDJ), ϕ_c, and meiosis II non-disjunction (MI NDJ), χ_c, are, respectively,

$$\phi_c = (2/3)[1-(-1/2)^{c+1}]$$
$$\chi_c = (2/3)[1-(-1/2)^{c}] \qquad [1]$$

where c=0,1,2,... (Mather, 1938; Côté and Edwards, 1975; Chakravarti and Slaugenhaupt, 1987). The behavior of these probabilities are shown in Figure 1, from which two features are evident. First, $\phi_c = 1-\chi_c/2$ for any number of chiasmata. Second, even though both ϕ_c and χ_c approach the value 2/3 by damped oscillations, they can both significantly exceed or fall short of 2/3, particularly if the number of crossovers is small.

For a genetic interval with an arbitrary crossover distribution, ϕ_c and χ_c are averaged over the probabilities of the numbers of chiasma to give the average probabilities of nonreduction as:

$$\phi = 1-y/2$$
$$\chi = y \qquad [2]$$

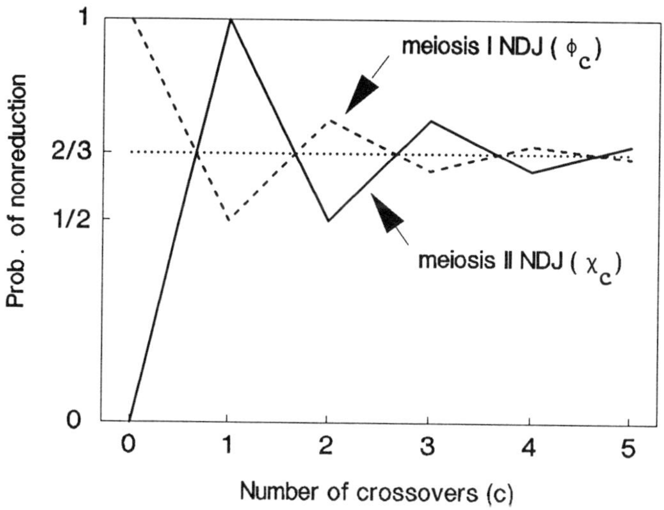

Figure 1: Probability of nonreduction as a function of the number of crossovers

(Sahar and Morton, 1986; Chakravarti and Slaugenhaupt, 1987), where y is the tetratype frequency, or, the probability of a recombinant meiotic tetrad.

Chakravarti and Slaugenhaupt (1987) have proposed that the tetratype frequency y be used as a linkage parameter, since y may be related to the gene-centromere map distance (w, in Morgans) and the recombination frequency (θ) by making varied assumptions regarding chiasma interference. Observe that when $\theta=0$ ($w=0$), $y=0$ so that $\phi=1$, $\chi=0$; also, when $\theta=1/2$ (c large, $w \rightarrow \infty$), $y=2/3$ so that $\phi=\chi=2/3$. For small map distances, there are either 0 or 1 crossovers with probabilities $1-2\theta$ and 2θ, respectively, so that $\theta=w$, $y=2\theta$; consequently, $\phi=1-\theta$ and $\chi=2\theta$. Thus, y may be viewed as a linkage parameter that varies between 0 and 2/3, although as Figure 1 suggests y may exceed 2/3 under strong interference (Ott et al., 1976; Chakravarti and Slaugenhaupt, 1987). The relationship between y, w and θ for three levels of interference is shown in Table 1. Numerical evaluations of these functions demonstrate that for map distances less than 20 cM, $y \approx 2\theta$ is a good approximation.

Table 1 : Relationship between y, w and θ

No. of chiasma (interference)	Tetratype Frequency	Mapping Function
$c \leq 1$ (complete)	$y=2\theta$	$w=\theta$
$c \leq 2$ (intermediate)	$y=3\theta-w$	Rao et al.(1977) (p=0.398)
c large	$y=(2/3)(1-e^{-3w})$	$w=-(1/2)\ln(1-2\theta)$

It would be useful to derive a general formula for y under any level of interference. This is not possible since a general chiasma distribution is unavailable. However, Ukai (1988) has suggested that any chiasma distribution may be approximated by a binomial distribution with one obligate chiasma. Whether such distributions are applicable to humans is, however, unknown.

For a genetic marker which is heterozygous in a host, the probability of nonreduction in an ovarian teratoma is simply given by ϕ and χ (equation [2]). The nonreduction probabilities are slightly complicated for autosomal trisomies or triploids since the genetic contribution of the parent in whom non-disjunction did not occur has to be considered. These probabilities depend on whether non-disjunction occurred at meiosis I or II, whether the error was maternal or paternal, as well as the marker genotypes of the parents. By enumerating all parental mating types and the trisomic offspring they produce, the probabilities of each trisomic genotype for each non-disjunction event can be calculated as functions of ϕ and χ. These probabilities as calculated by Chakravarti and Slaugenhaupt (1987) are shown in Table 2. Similar probabilities, in terms of the tetratype frequency y, are given in Sahar and Morton (1986).

Table 2. Probabilities of marker genotypes in trisomic offspring conditional on parental marker genotypes and a specific type of non-disjunction, where a, b, c and d are marker alleles. Reproduced from Chakravarti A & Slaugenhaupt SA (1987) with permission from Academic Press.

Mating type	Progeny type	Marker Genotype			Non-disjunction			
		Father	Mother	Child	♂I	♂II	♀I	♀II
A	1	aa	bb	aab	1	1	0	0
	2			abb	0	0	1	1
B	3	ab	ab	aaa,bbb	$\frac{1}{2}(1-\phi)$	$\frac{1}{2}(1-\chi)$	$\frac{1}{2}(1-\phi)$	$\frac{1}{2}(1-\chi)$
	4			aab,abb	$\frac{1}{2}(1+\phi)$	$\frac{1}{2}(1+\chi)$	$\frac{1}{2}(1+\phi)$	$\frac{1}{2}(1+\chi)$
C	5	ab	aa	aaa	$\frac{1}{2}(1-\phi)$	$\frac{1}{2}(1-\chi)$	$\frac{1}{2}$	$\frac{1}{2}$
	6			aab	ϕ	χ	$\frac{1}{2}$	$\frac{1}{2}$
	7			abb	$\frac{1}{2}(1-\phi)$	$\frac{1}{2}(1-\chi)$	0	0
D	8	aa	ab	aaa	$\frac{1}{2}$	$\frac{1}{2}$	$\frac{1}{2}(1-\phi)$	$\frac{1}{2}(1-\chi)$
	9			aab	$\frac{1}{2}$	$\frac{1}{2}$	ϕ	χ
	10			abb	0	0	$\frac{1}{2}(1-\phi)$	$\frac{1}{2}(1-\chi)$
E	11	ab	cc	aac,bbc	$1-\phi$	$1-\chi$	0	0
	12			acc,bcc	0	0	1	1
	13			abc	ϕ	χ	0	0
F	14	aa	bc	abb,acc	0	0	$1-\phi$	$1-\chi$
	15			aab,aac	1	1	0	0
	16			abc	0	0	ϕ	χ
G	17	ab	ac	aab	$\frac{1}{2}\phi$	$\frac{1}{2}\chi$	$\frac{1}{4}(1-\phi)$	$\frac{1}{4}(1-\chi)$
	18			abb,bbc	$\frac{1}{2}(1-\phi)$	$\frac{1}{2}(1-\chi)$	0	0
	19			abc	$\frac{1}{2}\phi$	$\frac{1}{2}\chi$	$\frac{1}{2}\phi$	$\frac{1}{2}\chi$
	20			acc,bcc	0	0	$\frac{1}{2}(1-\phi)$	$\frac{1}{2}(1-\chi)$
	21			aaa	$\frac{1}{4}(1-\phi)$	$\frac{1}{4}(1-\chi)$	$\frac{1}{4}(1-\phi)$	$\frac{1}{4}(1-\chi)$
	22			aac	$\frac{1}{4}(1-\phi)$	$\frac{1}{4}(1-\chi)$	$\frac{1}{2}\phi$	$\frac{1}{2}\chi$
H	23	ab	cd	abc,abd	ϕ	χ	0	0
	24			acd,bcd	0	0	ϕ	χ
	25			aa-,bb-	$1-\phi$	$1-\chi$	0	0
	26			-cc,-dd	0	0	$1-\phi$	$1-\chi$

A Theory for Multiple Markers

When multiple genetic markers on a single chromosome arm can be studied, observations on reduction or nonreduction on all markers can be simultaneously made. To assess the effect of recombination along the entire chromosome length, it would be useful to study gene-centromere and gene-gene map distances for all marked intervals. A theory for predicting the frequency of each type of unordered half-tetrad under recombination is thus necessary. There are two different approaches to this problem.

In the first method we assume that in the total genetic interval being studied there can be at most 3 chiasma with any particular marked interval sustaining at most 1 chiasma. This condition would be satisfied if the genetic map of a chromosome consisted of closely spaced markers. Then, under the assumption of 0, 1, 2, or 3 chiasma, one can enumerate all possible genotypes of non-disjunction gametes and calculate the probability of each unique type. For multiple exchanges, the relative probabilities of 2-, 3- or 4-strand exchanges may be calculated under the assumption of no chromatid interference, namely 1:2:1 and 1:6:9 for 2 and 3 chiasma, respectively. If the homologous chromosomes can be uniquely marked, say aaaaaaaa and bbbbbbbb, then each unordered half-tetrad can be classified into one of 12 genotypes depending on whether the two chromatids in the half-tetrad are crossover or non-crossover products. Any two crossover products in a half-tetrad may either be complementary or non-complementary. The conditional probabilities of these 12 types of half-tetrads under 0, 1, 2 or 3 chiasma and meiosis I non-disjunction are shown in Table 3. A specific type of tetrad is denoted as (i,j) where i and j are the numbers of crossovers on a particular chromatid, the * denoting complementary crossovers. Under meiosis II non-disjunction only 9 types of half-tetrads can be distinguished since complementary crossover products cannot be on sister chromatids, and these together with their respective probabilities are listed in Table 4.

Using the probabilities in Tables 3 and 4, the frequency of each half-tetrad can be computed in terms of the exchange probabilities for any gene-centromere interval. From observed frequencies of

Table 3: Probabilities of half-tetrads under MI NDJ with at most 3 chiasma

Class	Type	Conditional probability given c chiasma c=0	c=1	c=2	c=3
(0,0)	aaaaaaaaa bbbbbbbbb	1	1/4	1/16	1/64
(0,1)	aaaaaaaaa bbbbaaaaa		2/4	2/16	2/64
(0,2)	aaaaaaaaa bbbaaabbb			4/16	8/64
(0,3)	aaaaaaaaa bbaabbaaa				4/64
(1,1)*	aaaabbbbb bbbbaaaaa		1/4	2/16	3/64
(2,2)*	aabbbaaaa bbaaabbbb			1/16	3/64
(3,3)*	aabbaabbb bbaabbaaa				1/64
(1,1)	aaaaabbbb bbbaaaaaa			4/16	8/64
(2,2)	aabbbbbaa bbaaabbbb				8/64
(1,2)	aabbbaaaa bbbbbaaaa			2/16	16/64
(1,3)	aabbaabbb bbbbbbaaa				8/64
(2,3)	aabbaabbb bbaabbbbb				2/64

half-tetrads the exchange probabilities, P(c) for c = 0, 1, 2 or 3 chiasma , can then be estimated and used for calculating both the recombination value and map distance for a gene-centromere interval (Mather, 1938). Consequently, inter-marker map distances can also be calculated.

The second method may be more appropriate if a chromosome arm may sustain more than 3 chiasma and/or when the inter-marker distances are large enough that classifications of half-tetrads according to the types in Tables 3 and 4 may not be unequivocal. This method is particularly useful since gene-gene and gene-centromere distances may be independently estimated, in

Table 4: Probabilities of half-tetrads under MII NDJ with at most 3 chiasma

Class	Type	Conditional probability given c chiasma c=0	c=1	c=2	c=3
(0,0)	aaaaaaaaa aaaaaaaaa	1			
(0,1)	aaaaaaaaa aaaabbbbb		1	1/4	1/16
(0,2)	aaaaaaaaa aaabbbaaa			1/4	1/16
(0,3)	aaaaaaaaa aabbaabbb				2/16
(1,1)	aaaaabbbb aaabbbbbb			1/4	1/16
(2,2)	aabbaaaaa aaaabbaaa				1/16
(1,2)	aaaaaabbb aaabbbaaa			1/4	8/16
(1,3)	aaaaabbbb aabbbaabb				1/16
(2,3)	aaaaabbaa aabbbaabb				1/16

contrast with the first method where the successively larger gene-centromere intervals are correlated.

The second method assumes that reduction of a distal marker locus depends only on the reduction or nonreduction of its immediate proximal marker locus. This is equivalent to treating the proximal marker as a pseudocentromere relative to the distal marker. The calculations are exact only in the absence of interference. Consider a chromosome arm with two genetic markers A and B with linkage parameters as shown:

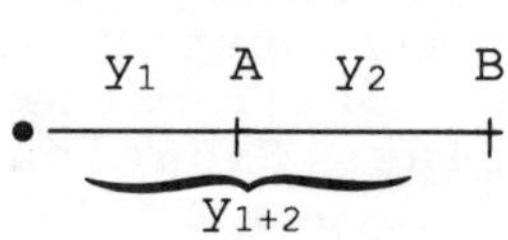

Note that the tetratype frequencies y_1 and y_{1+2} may be related to θ and w, and are gene-centromere distances. On the other hand, y_2 is a gene-gene distance but may be interpreted as a tetratype frequency assuming that marker locus A is a pseudocentromere. We denote homozygosity and heterozygosity by - and +, respectively. Then,

$$\text{Prob}\ \{\ B=+\ |\ A=+\ \} = 1-y_2/2, \qquad [3]$$

and,

$$\text{Prob}\ \{\ B=+\ |\ A=-\ \} = y_2. \qquad [4]$$

From elementary probability considerations, Prob $\{A=+, B=+\} = (1-y_2/2)P(A=+)$, and so on for the three other types of half-tetrads. Further specifications of $P(A=+)$ depend on whether nondisjunction occurred at meiosis I or II; under meiosis I and II NDJ, $P(A=+) = 1-y_1/2$ and y_1, respectively. Thus, the probabilities of the four types of half-tetrads are as provided in Table 5, using equations [2-4].

Table 5: Probabilities of half-tetrads under MI or MII NDJ

Half-tetrad	Probabilities under	
Genotype (A,B)	MI NDJ	MII NDJ
+ +	$(1-y_1/2)(1-y_2/2)$	$(y_1)(1-y_2/2)$
+ -	$(1-y_1/2)(y_2/2)$	$(y_1)(y_2/2)$
- +	$(y_1/2)(y_2)$	$(1-y_1)(y_2)$
- -	$(y_1/2)(1-y_2)$	$(1-y_1)(1-y_2)$

Under this model, the probability of nonreduction at the B locus is, under MII NDJ,:

$$\begin{aligned} \chi_{1+2} &= y_1(1-y_2/2) + (1-y_1)y_2 \\ &= y_1 + y_2 - (3/2)y_1y_2, \end{aligned} \qquad [5]$$

and, under MI NDJ the nonreduction probability is:

$$\begin{aligned} \phi_{1+2} &= (1-y_1/2)(1-y_2/2) + (1/2)y_1y_2 \\ &= 1 - \chi_{1+2}/2. \end{aligned} \qquad [6]$$

The proposed model thus preserves the relationship between ϕ and χ as in equation [2] and provides a useful combination formula for the tetratype frequency (equation [5]), since by definition $y_{1+2} = \chi_{1+2}$. For more than two markers the extension is straightforward. For k marker loci arranged as centromere-A_1-A_2-...-A_k, the probability of any half-tetrad is

$$\text{Prob}\{ A_1=\pm, A_2=\pm, \ldots, A_k=\pm \} = \left(\prod_{i=1}^{k-1} \text{Prob}\{A_{i+1}=\pm \mid A_i=\pm\}\right) P(A_1=\pm), \qquad [7]$$

where $P(A_1=\pm)$ depends on MI or MII NDJ and is a gene-centromere distance (equation [2]), and, $\text{Prob}\{A_{i+1}=\pm \mid A_i=\pm\}$ is a gene-gene distance (equations [3-4]). The combination formula for any length from interval i to j is,

$$y_{i\to j} = y_i + y_{i+1\to j} - (3/2) y_i \, y_{i+1\to j}$$

which can be evaluated by making repeated use of equation [5].

The advantages of this method are the use of any set of genetic markers irrespective of their inter-marker distances, the ability to include missing or uninformative marker loci, and ready application to trisomies and triploids where the genetic contribution of the other parent is considered independently for each interval.

THEORETICAL CONSIDERATIONS

Availability of a Genetic Map

To assess the role of recombination in non-disjunction for a particular chromosome it is necessary to have a detailed linkage map. Preferably such a map has multiple, closely spaced highly polymorphic genetic markers. Since recombination values are often different between the two sexes and vary along the chromosome for similar physical distances, sex-specific genetic maps together with physical localizations of markers are necessary. Fortunately, recombinant DNA methods have allowed us to proceed rapidly towards this goal. Comprehensive genetic maps for several human chromosomes are now available (White et al, 1986; Donis-Keller et al., 1987). Chromosome 21 in humans, due to its involvement in Down Syndrome, is of particular interest. Two groups have now constructed detailed maps of chromosome 21 consisting of over 20 loci (Tanzi et al., 1988; Warren et al., 1989).

Genetic maps are usually constructed without regard to the position of the centromere. The availability of DNA polymorphisms which are presumably at centromeres (Willard et al. 1986) will then allow alignment of any genetic map with respect to this landmark and studies of recombination in the pericentromeric region (Mahtani and Willard, 1988). Gene-centromere mapping in teratomas and trisomies will be particularly useful for this endeavor.

Choice of Genetic Markers

Non-disjunction of chromosomes at meiosis I or II can be inferred by studying the centromeric genotype which is nonreduced (heterozygous) or reduced (homozygous) under MI or MII NDJ, respectively. In humans, potentially useful centromeric markers may be either the classical centromeric heteromorphisms revealed by C- and Q-banding (Jacobs, 1977) or the DNA polymorphisms associated with α- satellite repeats (Jabs et al., 1984; Willard et al., 1986) found at the centromeres of all human chromosomes. For assessing the origin of non-disjunction the DNA polymorphisms are to be preferred since scoring individuals as homozygotes or heterozygotes at the centromere is objective and unequivocal. It is also clear that polymorphisms with multiple alleles each with approximately the same frequency are to be used since this maximizes the heterozygosity at the centromeric marker locus in the host. The same consideration holds for the choice of non-centromeric marker loci to be used for gene-centromere mapping. A marker locus with k codominant alleles $A_1, A_2, \ldots, A_k$ with frequency 1/k each has heterozygosity (1-1/k) with numerical values 0.50, 0.67, 0.75, 0.80 and 0.90 when k=2, 3, 4, 5 and 10, respectively. Several human polymorphic DNA markers have heterozygosities of 50% or more (Pearson et al., 1987), with those polymorphic for a variable number of tandem repeats (Nakamura et al., 1987) often having 80% or greater heterozygosity.

These considerations are sufficient for analyzing ovarian teratomas which arise in a single host. Trisomies and triploids also contain a chromosome complement transmitted by the parent in whom non-disjunction did not occur. This parent's genotype is crucial for determining the parental origin and meiotic stage of non-disjunction. A theoretical study of this problem has recently been made by Chakravarti (1989).

Given the genotypes of a trisomy and his/her parents, any genetic marker is useful for establishing the parental origin of non-disjunction, but only a centromeric marker is capable of establishing whether non-disjunction occurred at meiosis I or II. The probabilities of these events have been calculated by Chakravarti (1989), who also demonstrates the increased efficiency of establishing mode of origin of non-disjunction if parental origin is known. Consider a codominant marker locus with k equi-frequent alleles as before. Then, the probabilities of detecting meiotic stage, parental origin, and meiotic stage given parental origin are, respectively,

$$P(cen) = (k-1)\{(k-2)^2 + (2k-3)u\}/k^3$$
$$P(par) = (k-1)\{(k^2-3k+3)+(2k-3)[2u+y(1-3u)]/2\}/k^3$$
$$p(cen|par) = (k-1)\{(k^2-2) + u\}/k^3, \qquad [8]$$

where u is the probability of a meiosis II error and y is the tetratype frequency. Numerical values of these probabilities for various values of k and assuming u=0.2 (Hassold and Jacobs, 1984), and y=0 or 2/3 are provided in Table 6. For comparison, the heterozygosity (1-1/k) relevant for ovarian teratomas is also provided.

Table 6: Probabilities for detecting the origin of NDJ

No. of alleles	Teratomas: Heterozygosity	Trisomies: P(cen)	Trisomies: P(par) y=0	Trisomies: P(par) y=2/3	Trisomies: P(cen|par)	Trisomies: P(inf)
2	0.500	0.025	0.150	0.167	0.275	0.250
3	0.667	0.119	0.267	0.296	0.533	0.519
4	0.750	0.234	0.375	0.406	0.666	0.656
5	0.800	0.333	0.461	0.491	0.742	0.736
10	0.900	0.607	0.688	0.708	0.884	0.882

This table clearly shows that meiotic stage of non-disjunction can be inferred with much greater probability for teratomas (heterozygosity) than for trisomies (P(cen)) using the same genetic marker. For trisomies, meiotic stage of non-disjunction cannot be detected with high probability unless the centromeric marker is extremely polymorphic. The probability of detecting parental origin, P(par), depends on the degree of polymorphism, the frequency of meiosis II errors (u) and the distance between the marker and the centromere (y). As long as u<1/3, this probability increases as y increases, but as shown in Table 6, it does so slowly. In any case, the probability of establishing parental origin is always higher than establishing meiotic stage of non-disjunction. Finally, Table 6 demonstrates that once parental origin is known, the meiotic stage can be determined with higher efficiency (P(cen|par)). Formulas for arbitrary marker locus frequencies are provided in Chakravarti (1989).

The usefulness of a noncentromeric marker for detecting gene-centromere distances, P(inf), is a related problem answered by Chakravarti and Slaugenhaupt (1987). For ovarian teratomas it is simply the heterozygosity. For trisomies, the theoretical formula is given in Chakravarti and Slaugenhaupt (1987); values of P(inf) are also reported in Table 6. These values are useful for choosing the most informative marker loci for analysis.

Detection and Estimation of Gene-Centromere Linkage

The null hypothesis of no linkage between a gene and the centromere is specified by y=2/3. The evidence for linkage and a test of the hypothesis H_0 : y=2/3 is provided by the lod score method (Morton, 1955) commonly used in human family studies of linkage. The lod score at an assumed tetratype frequency y ($0 \leq y \leq 2/3$) is given by,

$$Z(y) = \log_{10}\{L(y)/L(2/3)\}, \qquad [9]$$

where L(y) is the likelihood of observations at a particular y value. Values of Z(y) can be computed at several assumed values of y to produce a standard lod table, from which the maximum likelihood estimate of y, $\hat{y}$, is calculated as that y value which maximizes Z(y) to yield $\hat{Z}=Z(\hat{y})$. Z(y) being a log likelihood may be added over independent observations at each calculated y value, and has the property (Morton, 1955),

$$\text{Prob}\{Z(\hat{y}) > \log_{10}A \mid y=2/3\} < A^{-1}. \qquad [10]$$

Morton (1955) suggested that $\log_{10}A = 3$, equivalent to odds of 1000:1, be used as a criterion to reject H_0. By convention, proof of linkage is obtained when $\hat{Z} > 3$, while absence of linkage at a particular y value is accepted if $Z(y) \leq -2$; otherwise, a decision is deferred and more data are collected.

For ovarian teratomas, tests of linkage and estimation of y is straightforward. Suppose that we have observations on m type I (meiosis I NDJ) and n type II (meiosis II NDJ) ovarian teratomas. At a particular marker locus, the number of nonreduced teratomas are m_1 for type I ($m_2 = m-m_1$)

and n_1 for type II ($n_2 = n - n_1$) teratomas. Then, the $\log_{10}$ likelihood of the total data is,

$$\log_{10}L(y) = \text{constant} + m_1\log(2-y) + (m_2+n_1)\log y + n_2\log(1-y). \quad [11]$$

The maximum likelihood estimate (MLE) of $y(\hat{y})$ is given by the solution to:

$$(m+n)y^2 - [3(m+n) - (2m_1+n_2)]y + 2(m_2+n_1) = 0 \quad [12]$$

with variance,

$$V(\hat{y}) = y(1-y)(2-y)/[n+(m+n)(1-y)]. \quad [13]$$

Estimation and tests of linkage may be performed separately for type I and II teratomas. From equations [12] and [13], the MLE of y and its variance from type I teratomas is,

$$\hat{y} = 2m_2/m, \quad V(\hat{y}) = y(2-y)/m,$$

and are obtained by setting $n_1 = n_2 = 0$. Analogously, for type II teratomas, on setting $m_1 = m_2 = 0$, we obtain (Ott, 1979),

$$\hat{y} = n_1/n, \quad V(\hat{y}) = y(1-y)/n.$$

For calculations of lod scores, we use the formula $Z(y) = \log_{10}L(y) - \log_{10}L(2/3)$ and equation [11].

When multiple markers are studied, given a specific gene order in the genetic map, one can use equation [7] for calculating the likelihood for an entire half-tetrad. As Table 5 demonstrates, for each genetic interval, the $\log_{10}L(y)$ value will be given by equation [11] and the total $\log_{10}$ likelihood can be written as a sum of $\log_{10}$ likelihoods for each interval. For each interval, then, a MLE of y, its variance and the associated lod score can be simply computed. Thus tests of recombination on each interval may be carried out separately. Tests of alternative gene orders can also be performed by comparing the

$\log_{10}L(y)$ values obtained under each locus order. Finally, the $\hat{y}$ value may be transformed to $\hat{\theta}$ and $\hat{w}$ values using one of the mapping functions in Table 1.

As mentioned earlier, under some circumstances, half-tetrads may also be classified according to the constituent chromatids being crossover or not, and the number of crossovers per chromatid (Tables 3 and 4). The analyses are identical, however, the calculation of the likelihood is not. Assume that k genetic markers are arranged as centromere-M_1-M_2-...-M_k with the ith distance being y_i (i=1,2,...k). The jth type of half-tetrad (Tables 3 and 4) has log likelihood,

$$\log_{10}L = \log_{10} \sum_{j} L(j|c)P(c) \qquad [14]$$

where c is the number of chiasma (c=0,1,2,3) and $L(j|c)$ is the conditional likelihood of the jth half-tetrad given c chiasma. In equation [14], $L(j|c)$ are numerical constants so that $\log_{10}L$ is a function of the exchange probabilities $P(c)$. Standard maximum likelihood methods (Rao, 1974) may be used to estimate the $P(c)$'s subject to the constraint $\sum_{c=0}^{3} P(c) = 1$. For the total interval, centromere to M_k, $\theta = [1-P(0)]/2$ and $w = [P(1) + 2P(2) + 3P(3)]/2$ (Mather, 1938), from which y may be estimated (Table 1). The map distances for each interval and consequently y_i (i=1,2,...,k) may be estimated by repeating the above analysis on the data but removing the most distal locus each time.

For autosomal trisomies, calculation of the log likelihood even for a single marker locus requires knowledge of the genetic contribution by the parent with normal disjunction. Then, $\log_{10}L(y)$ values may be calculated using equation [11] if the genetic contribution can be uniquely identified. However, y values should be calculated separately for meiosis I and II errors, by estimating ϕ and χ. These data can also be used to simultaneously estimate the recombination value on the normally disjoined chromosomes as outlined by Halloran and Chakravarti (1987).

The major problem with human data is that individual teratomas or trisomies may be uninformative for a particular marker locus, and that for trisomies the genetic contribution of the parent with normal disjunction may not be unique. In the latter case, for the gene-centromere interval, $\log_{10}L$ is calculated using equation [14] with c representing a particular non-disjunction type, i.e., maternal or paternal and meiosis I or II errors. The relevant conditional probabilities are those presented in Table 2. For any other gene-gene interval, we calculate likelihoods assuming the proximal locus as a pseudo-centromere (Morton et al., 1988) and using the conditional probability approach outlined in equation [7]. Once again, the specific probabilities are those in Table 2. For uninformative markers in teratomas and trisomies we calculate the likelihood as a sum of the likelihood of each type compatible with the observations.

Sample Size Requirements for Gene-Centromere Mapping

As mentioned earlier, gene-centromere linkage studies can be performed using lod scores and the method of maximum likelihood. To obtain significant evidence for linkage, we require the lod score to exceed 3. Using this criterion, we can calculate the sample size (n) of ovarian teratomas necessary to achieve a lod score of 3, i.e., $E[Z(y)]=3$ from equation [9] where E denotes the expected value. For a specified tetratype frequency y, we calculate the sample size by solving,

$$3 = E[\log_{10}L(y)] - \log_{10}L(2/3)$$

where $E[\log_{10}L(y)]$ is calculated using equation [11] but by replacing m_1, m_2, n_1 and n_2 by their expected values $m(1-y/2)$, $my/2$, ny and $n(1-y)$, respectively. The sample size for type I and type II ovarian teratomas is obtained by setting n=0 and m=0, respectively, and denoted by the generic symbol n. For autosomal trisomies, the calculations are complicated by the occurrence of several parental mating types. The calculations are similar and are outlined in detail in Chakravarti and Slaugenhaupt (1987). These sample sizes, for ovarian teratomas and autosomal trisomies, are graphically shown in Figure 2. This figure also gives the sample size necessary to achieve a lod score of 1.2 which corresponds to a significance level of 1%. This may be a more appropriate value since for gene-centromere mapping, the genetic markers studied must be known to be on the chromosome of interest.

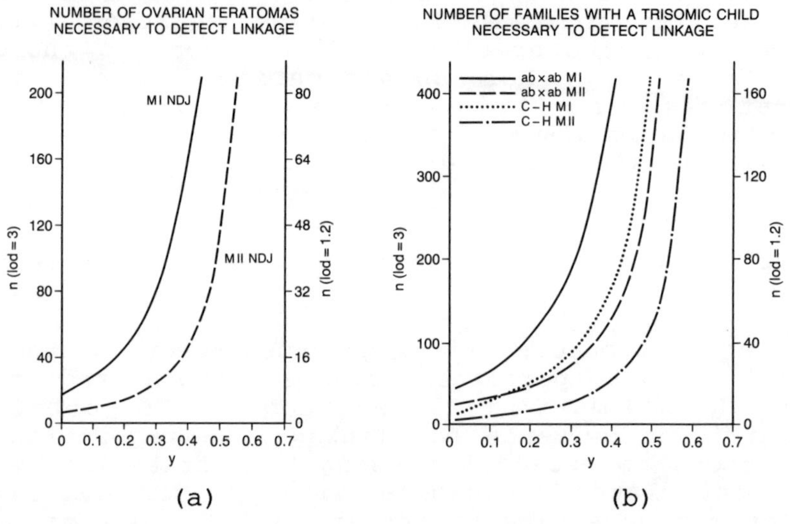

Figure 2: Sample size necessary to detect gene-centromere linkage (Fig. 2b reproduced from Chakravarti A & Slaugenhaupt SA (1987) with permission from Academic Press.)

Figure 2 demonstrates that detecting linkage using meiosis II errors is considerably more efficient than that obtained from meiosis I errors. Also, if genetic markers were spaced ≈ 20 cM apart (y ≈ 0.4) 170 type I ovarian teratomas and 50 type II teratomas are necessary to obtain a lod score of 3; the corresponding numbers for a 1.2 lod score are 68 and 20, respectively. These numbers are approximately the same for trisomies provided only the highly informative parental mating types C-H (Table 2) are utilized. These calculations emphasize the importance of having a fine structure map with markers less than 20 cM apart.

APPLICATIONS TO OVARIAN TERATOMAS AND TRISOMIES

Non-human Studies

Gene-centromere mapping has only recently been applied to human data. However these methods have been commonly used for linkage studies in experimental organisms by the use of artificial gynogenesis. The pioneering studies in this area were by Nace et al. (1970) who conducted gene-centromere linkage studies in *Rana pipiens*. In fish, gynogenones due to failure of meiosis II may be induced by UV light and have been extensively used to map genetic markers in zebrafish (Streisinger et al., 1986) and rainbow trout (Allendorf et al., 1986). Similar methods have also been used by Fuyama (1986) for studying the mechanisms of parthenogenesis in *Drosophila melanogaster*.

For humans, the closest experimental model for ovarian teratomas is provided by the mouse inbred strain LT/Sv (Eppig and Eicher, 1983), in which teratomas due to meiosis II non-disjunction occur spontaneously in ≈ 50% of females. These authors have demonstrated that six loci each known to be at least 30 cM from the centromere are indeed unlinked to the centromere in ovarian teratomas. Thus, recombination is not inhibited in mouse oocytes that later become teratomas, but this would not have been expected for meiosis II errors. This study also included two loci, Mpi-1 and Mod-1, known to be on mouse chromosome 9, but both unlinked to the centromere. Eppig and Eicher (1983) estimate the gene-gene recombination frequency between these as 22.9±5.4%, which agrees well with the frequency estimated from backcross data (18.6±3.3%). The frequency from ovarian teratomas was estimated assuming at most 2 crossovers. In contrast, we used the method outlined in Table 5 and assumed no interference to obtain the estimate 15.6±4.0%, a value closer to the backcross data. This example emphasizes the fact that gene-gene intervals may be reliably estimated even when neither marker is linked to the centromere.

In 1964, Merriam and Frost published studies of non-disjunction of the X chromosome in female *Drosophila melanogaster*. They studied 94 exceptional progeny arising from meiosis I non-disjunction and their genotypes at 5 X-linked genetic markers. Their data are useful for comparing the map distances estimated by two different gene-centromere linkage methods. In Table 2 of Merriam and Frost (1964), each of the 94 exceptional progeny have been classified according to the crossover types of each constituent chromosome as outlined in Table 3 of this paper. Thus, equation [14] may be used for estimating the exchange probabilities by the maximum likelihood method. These probabilities, as estimated for the five gene-centromere intervals, are provided in Table 7. The table also gives the control exchange probabilities for each gene-centromere interval. The control values for the largest interval, cen-sc, are given in Merriam and Frost (1964). The control values for the smaller gene-centromere intervals were calculated from the known control inter-marker map distances assuming the Carter-Falconer map function and the formulas of Morton and MacLean (1984). Under the assumption of at most two chiasma, Morton and MacLean (1984) showed that

Table 7: Exchange probabilities for meiosis I NDJ on D. melanogaster X chromosome (after Merriam & Frost, 1964). ( ) = control values

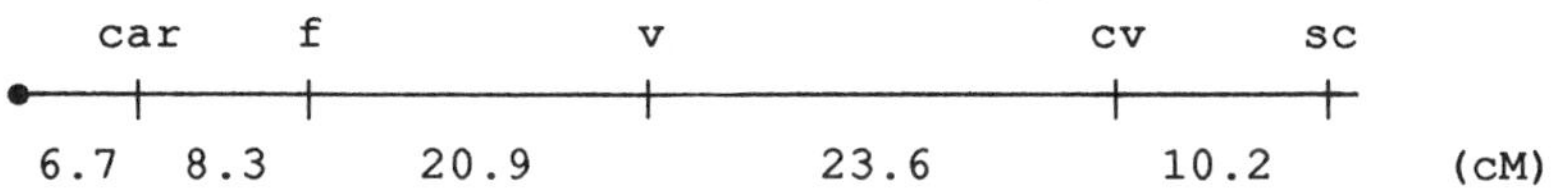

No. of chiasma	Exchange Probabilities				
	cen-car	cen-f	cen-v	cen-cv	cen-sc
0	.887(.886)	.745(.700)	.366(.316)	.305(.074)	.260(.046)
1	.113(.134)	.255(.299)	.619(.650)	.356(.663)	.245(.657)
2	--	(.001)	.015(.034)	.339(.263)	.476(.287)
3	--	--	--	--	.019(.101)
θ	.057(.127)	.127(.150)	.317(.342)	.347(.463)	.370(.477)
w	.057(.127)	.128(.151)	.325(.359)	.517(.595)	.627(.631)

P(c) = 1-2θ, 2(2θ-w) and 2(w-θ) where c=0,1 or 2 chiasma under any level of interference. The estimated standard errors for the various values in Table 7 are between 3 and 7.5%.

Table 7 is useful for making two major points. First, for chromosomes undergoing non-disjunction the estimated w and θ values are not significantly different from control values, but are consistently smaller. However, comparison of the exchange probabilities, particularly for the cen-cv and cen-sc interval, make it clear that there are more achiasmate and double chiasmate chromosomes and fewer single chiasmate chromosomes. Thus, there are alterations in the distribution of recombination events which are not evident on comparing θ values alone. However, most of this phenomenon is restricted to the terminal cv-sc interval.

Alternatively, the data may also be analyzed by studying reduction or nonreduction for the gene-centromere and four gene-gene intervals as outlined in Table 5 and equation [12]. These results are shown in Table 8 where the map distances have been calculated assuming complete (p=0) and no interference (p=1). The range of estimated w values agrees well with the estimated values in Table 7 and the control values, in spite of the lack of knowledge of the precise level of interference.

Table 8: Map distances for meiosis I NDJ on *Drosophila melanogaster* X chromosome (after Merriam and Frost, 1964)

Interval	Map distances (cM)			
	Equation [12]		Table 7	Control
	p=0	p=1		
gene-centromere				
cen-car	8.5	9.8	5.7	-
gene-gene				
car-f	8.7	10.1	7.1	8.3
f-v	15.6	20.9	19.7	20.9
v-cv	18.2	26.4	19.2	23.6
cv-sc	10.4	12.5	11.0	10.2

Human Studies

Linder (1969) originally proposed human ovarian teratomas to arise by meiosis II errors, but they are now thought to arise by diverse mechanisms (Carritt et al., 1983; Parrington et al., 1984; Surti et al., 1988). Of the five postulated mechanisms (Surti et al., 1988) only those arising by meiosis I and II errors are useful for gene-centromere mapping. We (U. Surti, R.E. Ferrell, A. Chakravarti) have recently begun a comprehensive study to determine the mode of origin of teratomas and to study gene-centromere linkage. The five modes of origin of ovarian teratomas can be distinguished with a combination of centromeric, genomic and single locus marker genotypes as shown in Table 9. The centromeric genotype refers to reduction or not at all informative centromeres and are assessed using

Table 9: Modes of Origin and Postulated Genotypes of Human Ovarian Teratomas (I/NI = identical/not identical to host; +/- = heterozygous/homozygous)

Type	Mechanism	Teratoma Genotype: Centromeric	Genomic	Marker
I	Meiosis I error	+	NI	+,-
II	Meiosis II error	-	NI	+,-
III	Endoreduplication	-	NI	-
IV	No meiosis	+	I	+
V	Fusion of two Ova	+,-	NI or I	+,-

chromosomal heteromorphisms or α-satellite polymorphisms at centromeres. The marker loci refer to markers studied on different chromosomes that can undergo reduction and are studied using DNA polymorphisms. The genomic genotype refers to assessment using mini-satellite DNA probes (Jeffreys et al., 1985) or the M13 probe (Vassart et al., 1987) to detect if the teratoma is identical to the host or not. Non-identity of the host and teratoma genomic genotype is confirmed by missing fragments in types I-III; for type V this occurs with 50% probability. However, this test is equivocal because these probes do not sample the whole genome.

In our current study we have observations on 32 ovarian teratomas among which no type V have been observed. Among these, the numbers of types I, II, III, IV, I or IV, II or III are 9, 10, 0, 3, 1 and 9, respectively. Maximum likelihood estimation from these data demonstrate that 62±9% ~ 90±5% of these teratomas are informative for linkage (types I and II). The range of values is obtained by assuming that none or all of the type II or III teratomas are type II. Thus, a major fraction of ovarian teratomas will be informative for linkage.

Our preliminary studies have also utilized four DNA polymorphisms on human chromosomes 1 and 13. These data are shown in Table 10, and calculations using equations [12-13] on these limited data demonstrate that all of these genetic

Table 10: Gene-Centromere distances in Human Ovarian teratomas (+/- = heterozygous/homozygous)

Chromosome:		1p	13q		
Locus:		D1S57	RB	D13S4	D13S3
Meiosis I NDJ	+	5	0	3	0
	-	3	6	1	6
Meiosis II NDJ	+	6	3	4	2
	-	1	3	0	5
Sample Size		15	12	8	13
$\hat{y} \pm$ std. dev.		.84±.13	.75±.16	1.00±.00	.62±.17

markers are unlinked to the centromere. The three loci on chromosome 13 are arranged as centromere-RB-D13S4-D13S3. Four ovarian teratomas were informative for all three of these genetic markers. Among these two were type I and had genotype -+-; these arose by at least three crossover events. The two arising by meiosis II errors had genotypes -+- and ++- and were thus both due to 2 strand or 4 strand double crossovers. Therefore, the occurrence of recombination in ovarian teratomas is clearly demonstrated.

Studies of recombination in autosomal trisomies have recently been performed by Antonarakis et al. (1986), Warren et al. (1987), Stewart et al. (1988) and Hassold et al. (1988). The pioneering studies on liveborn trisomy 21 (Down Syndrome) by Antonarakis et al (1986) and Warren et al. (1987) demonstrated reduced recombination on the non-disjoined chromosomes but not on the normally disjoined chromosomes. However, studies on spontaneously aborted trisomy 21 cases (Stewart et al., 1988) and trisomy 13 cases (Hassold et al, 1988) do show the occurrence of recombination. Recently, we (S.E. Antonarakis, A.C. Warren, A. Chakravarti) have analyzed further family data to address this question.

At the present time 61 families, each containing a child with Down Syndrome, an unaffected sibling and their parents have been sampled. Ten polymorphisms using the DNA probes D21S13, D21S110, D21S1/D21S11, D21S8, APP, D21S111, D21S82, D21S3, D21S112 and COL6A1 were examined in these families. The DNA polymorphisms at these loci, the locus order and the map distances between them are summarized in Antonarakis et al. (this volume).

Among these families, 55 (90.2±3.8%) were of maternal origin, while 6 (9.8±3.8%) were of paternal origin. The frequency of paternal non-disjunction observed in this sample is significantly lower than the 20% estimated using chromosomal heteromorphisms (Hassold and Jacobs, 1984). Since scoring cytogenetic variants is not unequivocal, it is possible that the frequency of paternal non-disjunction was previously overestimated. Centromeric heteromorphisms were not studied in these 61 families and so we have no direct estimate of the frequency of meiosis I and II errors. However, D21S13 is the most centromeric marker and, at this locus the numbers of maternal I, maternal II, paternal I and paternal II non-disjunction were 13, 7, 0 and 2 (n=22), respectively. Once again, the proportions of maternal and paternal non-disjunction were 90.9% and 9.1%, respectively, while the proportion of meiosis I and II errors are 59.1% and 40.9%, respectively. This analysis assumes D21S13 not to recombined with the centromere. The probabilities of meiosis I and II errors have been previously estimated to be 80% and 20%, respectively (Hassold and Jacobs, 1984). The observation of 59.1% meiosis I errors is lower than, but not significantly different from, the expectation and provides preliminary evidence that D21S13 recombines with the centromere.

The studies of genetic markers clearly identified recombination events. Among the 61 families, 44 families appeared nonrecombinant; 14 families had one crossover and 3 families had two crossovers. Among the 6 families with paternal non-disjunction, 3 had no recombinants while 3 had a single crossover; thus, among the 55 maternal non-disjunction families, the number of families with 0, 1 or 2 crossovers was 41, 11 and 3, respectively. These observations suggest an increase in the number of families with no crossovers. However, these results should be viewed with great caution since not all genetic markers were informative in all families and thus recombination events may have gone undetected. A definitive answer can only be given once maximum likelihood analysis of the data are performed. We are in the process of completing such analyses on a set of 150 families.

ENVOI

The success of gene-centromere mapping in humans rests on two major advances. First, DNA polymorphisms at numerous chromosomal locations and at centromeres, as well as a fine structure genetic map of these markers, are necessary. Second, population genetics theory to predict the frequencies of non-disjunction products under recombination is required. Recent advances on both fronts now make gene-centromere mapping a feasible goal. The analysis of genetic marker data from ovarian teratomas, trisomies and triploids using the methods of gene-centromere mapping will provide valuable information on the role of recombination in the proper disjunction of chromosomes. It is expected that the ease of these studies in experimental organisms will be equally possible in humans.

ACKNOWLEDGEMENT

We thank Tara Cox for preparation of the manuscript. This study was partially supported by NIH grants GM33771 (AC, PPM), HD00774 (AC), CA43882 (US, REF, AC), HD19591 (SEA) and GM41015 (SEA).

REFERENCES

Allendorf FW, Seeb JE, Knudsen KL, Thorgaard GH, Leary RF (1986). Gene-centromere mapping of 25 loci in rainbow trout. J Hered 77:307-312.

Antonarakis SE, Chakravarti A, Warren AC, Slaugenhaupt SA, Wong C, Halloran SL, Metaxotou C (1986). Reduced recombination rate on chromosomes 21 that have undergone recombination. Cold Spring Harb Symp Quant Biol 51:185-190.

Baker B, Carpenter A, Esposito M, Esposito R, Sandler L (1976). The genetic control of meiosis. Annu Rev Genet 10:53-134.

Bridges CB (1916). Non-disjunction as proof of the chromosome theory of heredity. Genetics 1:1-52,107-163.

Carr DH, Gedeon M (1977). Population cytogenetics of human abortuses. In Hook EB, Porter IH (eds): "Population Cytogenetics," New York: Academic Press, pp 1-9.

Carritt B, Parrington J, Welch H, Povey S (1982). Diverse origins of multiple ovarian teratomas in a single individual. Proc Natl Acad Sci (USA) 79:7400-7404.

Chakravarti A (1989) The probability of detecting the origin of non-disjunction of autosomal trisomies. Am J Hum Genet (in press).

Chakravarti A, Slaugenhaupt SA (1987). Methods for studying recombination on chromosomes that undergo non-disjunction. Genomics 1:35-42.

Côté GB, Edwards JH (1975) Centromeric linkage in autosomal trisomies. Ann Hum Genet 39:51-59.

Darlington CD (1929). Chromosome behavior and structural hybridity in the Tradescantiae. J Genet 21:207-286.

Dobzhansky T (1933). Studies on chromosome conjugation. II. Z Indukt Abstamm Vererbungsl 65:269-309.

Donis-Keller H and 32 others (1987). A genetic linkage map of the human genome. Cell 51:319-337.

Eppig JT, Eicher M (1983). Application of the ovarian teratoma mapping method in the mouse. Genetics 103:797-812.

Fuyama Y (1986). Genetics of parthenogenesis in Drosophila melanogaster. I. The modes of diploidization in the gynogenesis induced by a male-sterile mutant, MS(3)K81. Genetics 112: 237-248.
Halloran SL, Chakravarti A (1987). DSLINK: A computer program for gene-centromere linkage analysis in families with a trisomic offspring. Am J Hum Genet 41:350-355.
Hassold TJ, Jacobs PA (1984). Trisomy in man. Annu Rev Genet 18:69-97.
Hassold TJ, Chiu D, Yamane J (1984). Parental origin of autosomal trisomies. Ann Hum Genet 48:129-144.
Hassold TJ, Jacobs PA, Leppert M, Sheldon M (1987). Cytogenetic and molecular studies of trisomy 13. J Med Genet 24:725-732.
Henderson SA (1970). The time and place of meiotic crossing-over. Annu Rev Genet 4:295-324.
Henderson SA, Edwards RG (1968). Chiasma frequency and maternal age in mammals. Nature 218:22-28.
Hoffner L, Surti U (1989). Cytogenetic findings in ovarian teratomas. (manuscript).
Jabs EW, Wolf SF, Migeon BR (1984). Characterization of a cloned DNA sequence that is present at the centromeres of all human autosomes and the X chromosome and shows polymorphism variation. Proc Natl Acad Sci (USA) 81:4884-4888.
Jacobs PA (1977). Human chromosome heteromorphisms (variants.) Prog Med Genet 2:252-274.
Jeffreys AJ, Wilson V, Thein SL (1985). Hypervariable "mini-satellite" regions in human DNA. Nature 314:67-73.
Kaiser-McCaw B, Hecht F, Linder D, Louvrien EW, Wyandt H, Bacon D, Clark B, Lea N (1976). Ovarian teratomas: cytologic data. Cytogenet Cell Genet 16:391-395.
Kajii T, Ohama K (1977). Androgenetic origins of hydatidiform moles. Nature 268:633-634.
Linder D (1969). Gene loss in human teratomas. Proc Natl Acad Sci (USA) 63:699-704.

Mahtani MM, Willard HF (1988). A primary genetic map of the pericentromeric region of the human X-chromosome. Genomics 2:294-301.

Mather K (1938). Crossing over. Biol Rev 13:252-292.

Maynard Smith J (1978). "The Evolution of Sex." New York: Cambridge Univ Press.

Merriam JR, Frost JN (1964). Exchange and non-disjunction in female Drosophila melanogaster. Genetics 49:109-122.

Morton NE (1955). Sequential tests for the detection of linkage. Am J Hum Genet 7:277-318.

Morton NE, MacLean CJ (1984). Multilocus recombination frequencies. Genet Res 44:99-108.

Morton NE, Wu D, Jacobs PA (1988). Origin of sex chromosome aneuploidy. Ann Hum Genet 52:85-92.

Nace GW, Richards CM, Asher JH (1970). Parthenogenesis and genetic variability. I. Linkage and inbreeding estimations in the frog, Rana pipiens. Genetics 66:349-368.

Nakamura Y, Leppert M, O'Connell O, Wolff R, Holm T, Culver M, Martin C, Fujimoto E, Hoff M, Kumlin E, White RL (1987). Variable number of tandem repeat (VNTR) markers for human gene mapping. Science 235:1616-1622.

Ohama K (1987). Androgenesis and parthenogenesis in humans. In Vogel F, Sperling K (eds): "Human Genetics," Berlin: Springer Verlag, pp 245-249.

Ott J (1979). Human gene mapping by postreduction and recombination frequencies under complete interference. Clin Genet 15:11-16.

Ott J, Linder D, McCaw BK, Louvrien E, Hecht F (1976). Estimating distances from the centromere by means of benign ovarian teratomas in man. Ann Hum Genet 40:191-196.

Parrington J, West L, Povey S (1984). The origin of ovarian teratomas. J Med Genet 21:4-12.

Pearson PL, Kidd KK, Willard HF (1987). Report of the Committee on Human Gene Mapping by Recombinant DNA techniques. Cytogenet Cell Genet 46:390-566.

Peterson WF, Prevost EC, Edmunds FT, Huntly JM, Morris FK (1955). Benign cystic teratomas of the ovary: A clinico-statistical study of 1007 cases with a review of the literature. Am J Ob Gyn 70:368-382.

Polani PE, Crolla JA (1982). Experiments on female mammalian meiosis. In Crosiganini PG, Rubin BL (eds): "Genetic control of gamete production and function," New York: Academic Press, pp 171-186.

Rao CR (1974) Linear Statistical Inference and its Applications. 2nd edition. New York: John Wiley.

Rao DC, Morton NE, Lindsten J, Hulten M, Yee S (1977). A mapping function for man. Hum Hered 27:99-104.

Sahar S, Morton NE (1986). Origin of teratomas and twins. Hum Genet 74:215-218.

Stewart GD, Hassold TJ, Berg A, Watkins P, Tanzi R, Kurnit DM (1988). Trisomy 21 (Down Syn drome): Studying non-disjunction and meiotic recombination by using cytogenetic and molecular polymorphisms that span chromosome 21. Am J Hum Genet 42:227-236.

Streisinger G, Singer F, Walker C, Knauber D, Dower N (1986). Segregation analysis and gene-centromere distances in zebrafish. Genetics 112:311-319.

Sugawara S, Mikamo K (1983). Absence of correlation between univalent formation and meiotic non-disjunction in aged female chinese hamsters. Cytogenet Cell Genet 35:34-40.

Surti U, Deka R, Hoffner L, Majumder PP, Chakravarti A, Ferrell RE (1988). Human ovarian teratomas: Origin and gene mapping using cytogenetic and DNA markers. Am J Hum Genet 43:A34.

Tanzi R, Haines J, Watkins P, Stewart G, Wallace M, Hallewell R, Wong C, Wexler N, Conneally P, Gusella J (1988). Genetic linkage map of human chromosome 21. Genomics 3:129-136.

Therman E (1986). "Human Chromosomes: Structure, Behavior, Effects." New York: Springer-Verlag.

Ukai Y (1988). Fitting of a binomial distribution to the number of chiasmata per bivalent. Jpn J Genet 63:259-265.

Vassart G, Georges M, Monsieur R, Brocas H, Lequarre AS, Christhophe D (1987). A sequence in M13 phage detects hypervariable minisatellites in human and animal DNA. Science 235:683-684.

Warren AC, Chakravarti A, Wong C, Slaugenhaupt SA, Halloran SL, Watkins PC, Metaxotou C, Antonarakis SE (1987). Evidence for reduced recombination on the non-disjoined chromosomes 21 in Down Syndrome. Science 237:652-654.

Warren AC, Slaugenhaupt SA, Lewis JG, Chakravarti A, Antonarakis SE (1989). A genetic linkage map of human chromosome 21. Genomics. (in press)

White RL, Leppert M, O'Connell P, Nakamura Y, Julier C, Woodward S, Silva A, Wolff R, Lathrop M, Lalouel J-M (1986). Construction of human genetic linkage maps. I. Progress and Perspectives. Cold Spring Harbor Symp Quant Biol 51:29-38.

Willard HF, Waye JS, Skolnick MH, Schwartz CE, Powers VE, England SB (1986). Restriction fragment length polymorphisms at the centromere of human chromosomes using chromosome specific alpha satellite DNA. Proc Natl Acad Sci (USA) 83:5611-5615.

Molecular and Cytogenetic Studies of
Non-Disjunction, pages 81–100

CYTOLOGICAL AND MOLECULAR STUDIES OF NUCLEOLAR ORGANIZING REGION VARIANTS AND RECOMBINATION IN TRISOMY 21

Diane Roulston, Stylianos E. Antonarakis, John G. Lewis, Maimon M. Cohen and Stuart Schwartz

Division of Human Genetics, Departments of Obstetrics/Gynecology and Pediatrics (DR, MMC, SS), University of Maryland School of Medicine, Baltimore, Maryland 21201

Genetics Unit, Department of Pediatrics (SEA, JGL) The Johns Hopkins University School of Medicine, Baltimore, Maryland 21205

INTRODUCTION

The etiology of nondisjunction is poorly understood, despite Down syndrome being the single most common cause of mental retardation and chromosome abnormalities being associated with over half of first trimester pregnancy losses. The maternal age effect, one of the initial risk factors described, implicated maternal factors in the cause of Down syndrome (Penrose, 1933). Since the cytological discovery of trisomy 21 (Lejeune, 1959), much attention has been directed to factors possibly influencing oogenesis, from events early in fetal development through effects that could be cumulative as females age (Hassold and Jacobs, 1984). However, early studies of chromosome structural heteromorphisms demonstrated that paternal nondisjunction also occurs, and that chromosomal malsegregation at both meiotic divisions contributes to the incidence of Down syndrome (Uchida, 1973; Lieznerski and Lindsten, 1972).

This study was supported in part by grants from the International Order of the Alhambra; the Bressler Research Foundation and Designated Research Initiative Funds, University of Maryland at Baltimore; and the Medical Biotechnology Center of the Maryland Biotechnology Institute.

Theories attempting to explain the greater prevalence of maternal meiotic errors, including advanced maternal age, often incorporate the differences in the process of gametogenesis between males and females (Hassold and Jacobs, 1984). The fact that human female oocytes remain in dictyotene of prophase I from fetal development only to complete meiosis after fertilization, sometime between puberty and menopause (Donahue, 1970) may be particularly important. Nucleolar persistence, for example, is one such mechanism proposed to disrupt orderly chromosome assortment (Polani, 1960). Similarly, satellite association between the acrocentric chromosomes, a common cytological phenomenon observed in both meiosis and mitosis, was found to be increased in parents of trisomy 21 offspring (Mattei, 1974). However, the possibility of an etiologic effect has not been supported by other investigators (Gould and Martin-DeLeon, 1987).

More recently, the presence of a double nucleolar organizing region (dNOR) has been suggested as an important etiologic factor in nondisjunction (Jackson-Cook et al, 1985). These variant structures, found on the acrocentric chromosomes (Fig. 1) were found to be strongly associated with parents of children with trisomy 21. The dNOR was not

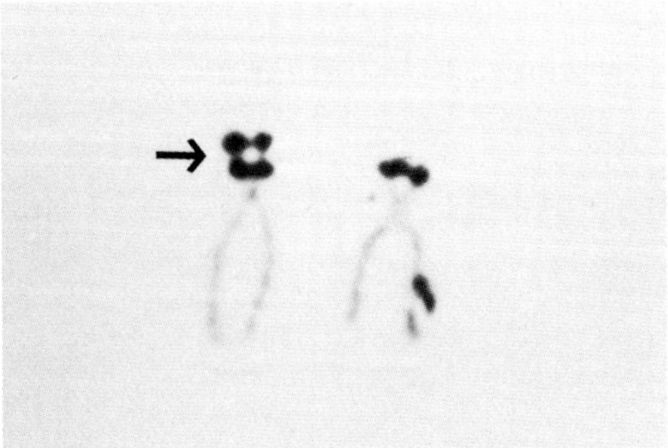

Figure 1. A double nucleolar organizing region (dNOR) variant chromosome (arrow)

always found on chromosome 21. However, in all 13 cases, the parent in whom nondisjunction occurred was found to also possess the dNOR's, thereby strengthening the association. While the original study did not detect any dNOR's in a control population, subsequent studies have done so (Spinner et al, 1986; Melnyk et al, 1987); however, these studies did not determine parental origin of the supernumerary chromosome in the Down syndrome parent population.

These proposed factors affecting nondisjunction have in common the concept that cellular mechanisms exist which may prevent chiasma formation, which would in turn cause misalignment and malsegregation of bivalents on the metaphase plate. Evidence from animal studies supports the tenet that recombination is required for proper chiasma formation (Rasmussen and Holm, 1985), and that failure of recombination may increase aneuploidy (Grell, 1985; Henderson and Edwards, 1968). Additionally, smaller chromosomes have a higher incidence in aneuploid spontaneous abortuses (Hassold et al, 1984), and chromosome length is known to correlate with chiasma frequency (Hulten, 1974). Hence, there is much speculation that cellular events early in prophase of meiosis I of oogenesis serve to decrease recombination and thereby lead to increased nondisjunction. The recent availability of specific molecular markers localized along the long arm of chromosome no. 21 allows this question to be addressed directly, by examining the inheritance patterns of genetic polymorphisms for recombination events on the nondisjoined chromosomes (Chakravarti and Slaugenhaupt, 1987). A recent study indicates lower than normal frequencies of recombination on nondisjoined chromosomes (Warren et al, 1987), thereby indirectly supporting a cytological mechanism.

The present investigation was undertaken with the aim of partitioning out the importance of various proposed etiologic factors by examining families with at least one trisomy 21 child. The frequency of recombination on the nondisjoined chromosomes and presence of dNOR's were determined and analyzed with respect to origin of nondisjunction and maternal age. The following questions were posed:

(1) Is there an association between the dNOR variant chromosome and the origin of nondisjunction in trisomy 21?

(2) Is there an association between nondisjunction and recombination on chromosome no. 21?

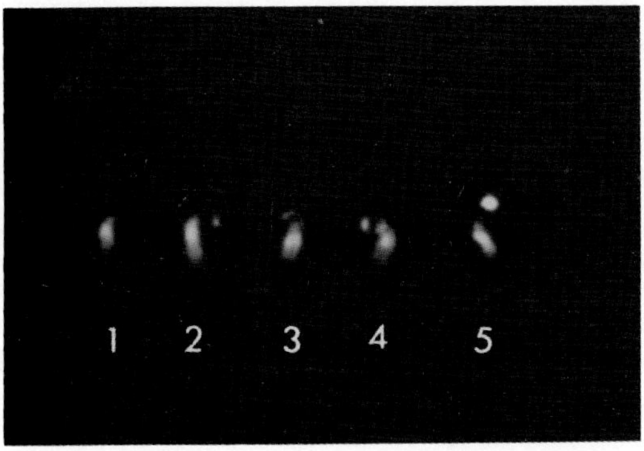

Figure 2. Quinacrine fluorescence heteromorphisms of chromosome no. 21. Intensity classes range from 1 (no fluorescence) to 5 (brilliant). The fourth chromosome also has a large p11.2 region (value = 5).

MATERIALS AND METHODS

Study Group

Forty couples with at least one child trisomic for chromosome no. 21 participated in the dNOR study by donating peripheral blood samples and completing a family history interview or questionnaire. Maternal age at the time of birth of the Down syndrome child was documented. Fourteen couples with no history of pregnancy loss and at least one normal child served as controls for parental dNOR frequency comparisons. Peripheral blood samples also were obtained from 20 Down syndrome children and their parents for chromosome heteromorphism and DNA analysis.

Chromosome Analysis

Cytological polymorphisms of the short arm and satellite region of chromosome no. 21 can be detected with quinacrine fluorescence and silver staining. This technique is well established for determining meiotic and parental origins of nondisjunction (Juberg and Mowery, 1983; Jacobs and Mayer, 1981; Mikkelsen et al, 1980). Fluorescence intensity of these chromosomal regions depends, at least in part, on the amount of satellite DNA present in the sub-bands (Gosden et al, 1981; Korenberg and Engels, 1978; Weisblum, 1973). As Figure 2 illustrates, size and quinacrine fluorescence intensity may be quantified for both the acrocentric satellites and proximal region of the short arm (p11.2). Comparison to internal constant bands leads to five classes of intensity (ISCN, 1985).

The amount of silver staining of the NOR regions appears to depend on the level of ribosomal RNA transcriptional activity of the previous cell division (Hubbell, 1985). While there is variable appearance of a particular NOR from cell to cell, the average size is a constant quantity, and transmitted in a Mendelian codominant fashion (Markovic et al, 1978). Hence, combining the two staining procedures allows the detection of 5 heritable structural quantities: NOR size, and

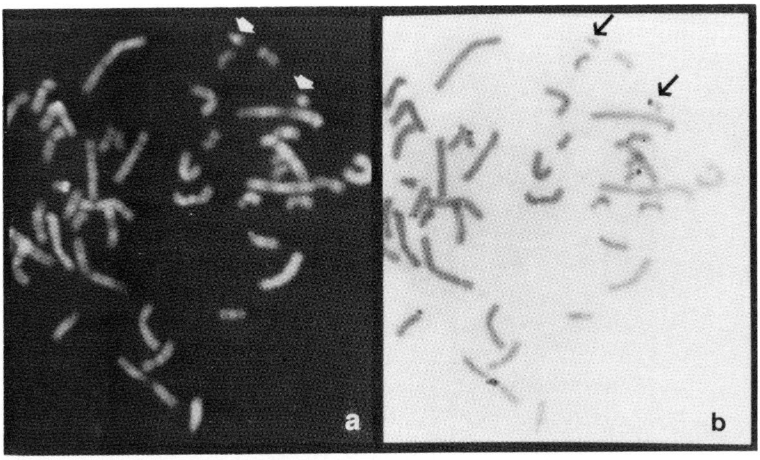

Figure 3. Sequential QFQ and AgNOR staining of the same metaphase cell; chromosome no. 21's are designated by arrows (a) QFQ (b) AgNOR.

quinacrine fluorescence intensity for both the satellites and the proximal short arm of chromosome 21. For example, using these methods, the chromosome 21 in Figure 3b shows complete absence of an NOR region and its homologue shows a very pale short arm and satellite staining in the Q-banded preparation (Fig. 3a).

PHA-simulated lymphocytes were cultured and harvested for chromosome analysis following the method of Ikeuchi (1984). Quinacrine fluorescence staining methods followed Caspersson et al (1971). At least ten metaphase cells were analyzed for chromosome 21 satellite and p arm values of size and intensity, and locations were recorded at the microscope. Two cells were photographed per person.

After destaining the Q-banded slides, the chromosomes were silver stained (Howell and Black, 1980) and the number 21's were relocated for each cell. Coded slides were made for a blinded analysis of dNOR frequencies, which included examination of all acrocentrics in at least 10 cells.

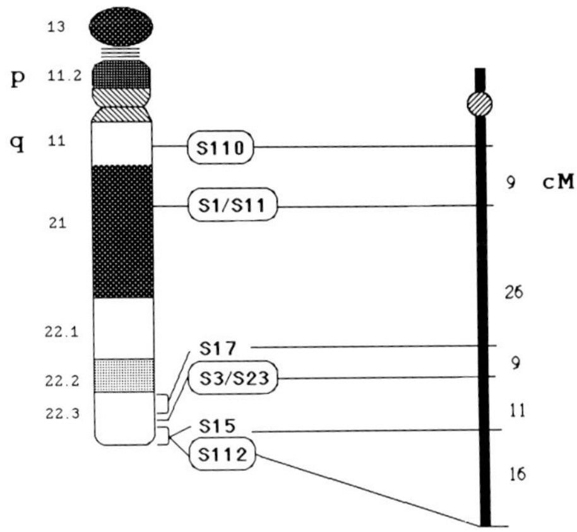

Figure 4. Composite map of DNA probe locations. The anonymous sequences shown encircled were used in this study. Map distances in centimorgans (cM) are per Warren et al (1989). The physical map (idiogram) shows localization within specific bands (Kaplan and Carritt, 1987).

DNA Markers

Figure 4 shows the identities of the six unique chromosome 21 DNA probes used in this study, their physical map locations and relative map distances where known. One RFLP (D21S110) is relatively close to the centromere, as it was obtained from screening genomic libraries for members of the pericentromeric "724" family of alpha satellite DNA, until a sequence was found which mapped uniquely to chromosome 21q11 (Kurnit et al, 1986; Kurnit et al, 1984). Two pairs of tightly linked RFLP's, D21S3/S23, and D21S1/S11, form very informative haplotypes, and have been well mapped for physical and genetic locations on the long arm of chromosome 21 (Watkins et al, 1985; Kazazian et al, 1985; Tanzi et al, 1988; Warren et al, 1989). D21S112 detects VNTR sequences, thereby allowing a limited form of DNA fingerprinting. This probe is highly informative for determining parental origin of chromosome no. 21 nondisjunction, and it is also the most telomeric locus in the study.

DNA was isolated from white blood cells (Kan et al, 1977) and Southern blots were prepared from restriction enzyme digests (Southern, 1975). The chromosome 21 specific DNA fragments were purified from their vectors, and labeled with high specific activity ^{32}P by oligonucleotide primer labeling methods (Feinberg and Vogelstein, 1983).

RESULTS

Parental Origin and Meiotic Errors

Combining the observations obtained from utilizing both cytological and molecular methods allowed determination of the origin of nondisjunction for either parent or stage of meiosis in all 20 Down syndrome children studied (Table 1). The distribution of the origin of nondisjunction agrees well with previously reported values from the literature (Hassold and Jacobs, 1984). Approximately 81% of all events were maternal, and 19% paternal in origin; 65% of the cases could be identified for both parent and meiotic stage of origin. Meiotic stage of nondisjunction was determined by cytogenetic markers in all cases, while parent of origin could be determined by either cytological markers (37.5%), molecular markers (37.5%), or both (25%).

Table 1. Origin of Nondisjunction by Parent and Meiotic Division

	MAT			PAT			MAT/PAT	
Meiosis	I	II	I/II	I	II	I/II	I	II
Number	10	1	2	0	2	1	3	1
	--	81%	--	--	19%	--		

Parental origin known = 80%
Meiotic division known = 85%
Parent and meiotic division known = 65.0%

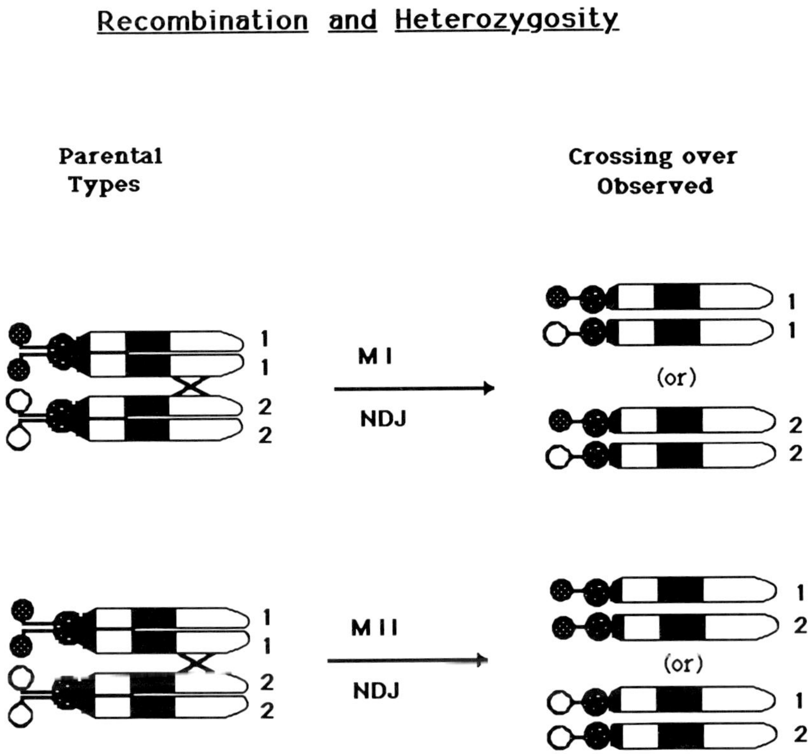

Figure 5. Detection of recombination between informative short arm satellite heteromorphisms [designated as open (o) or filled circles (•)], and telomeric loci (designated numerically as 1 or 2). In meiosis I NDJ, although a crossover may occur, recombination would not be detected by reduction to homozygosity 50% of the time (not shown).

Recombination

Recombination was detected in the long arm of chromosome 21 pairs undergoing nondisjunction in 5 of 9 informative cases. When the origin of nondisjunction can be determined by informative centromeric markers, recombination on the nondisjoined chromosomes can be

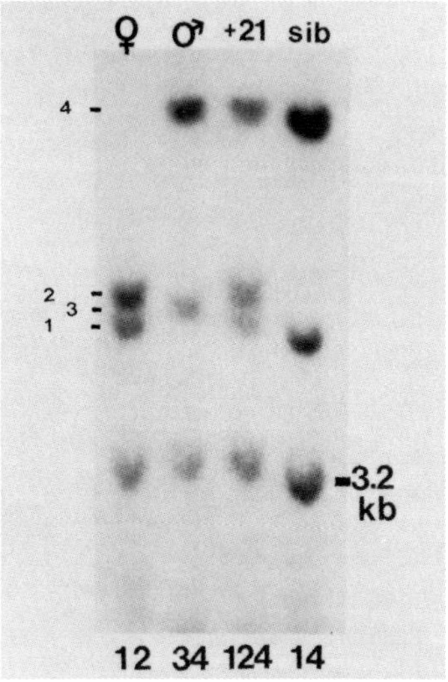

Figure 6. Partial autoradiogram of family DNA analysis. Rsa I digests were made of genomic DNA, run overnight on 1% agarose gels and transferred to nitrocellulose filters. Southern blots were hybridized with purified L427 inserts labeled with high specific activity ^{32}P. The 3.2 kb constant bands may be used as a reference point, and all bands larger than 3.2 kb are represented. Allele designations are specific within a family.

detected, according to the method illustrated in Figure 5. In a meiosis I event with no recombination, heterozygosity for the telomeric marker is retained. However, recombination is demonstrated if a marker which was heterozygous in the parent of origin, is found to be homozygous in the trisomic offspring. In a meiosis II event, without crossing-over, the trisomic offspring would be homozygous for the alleles. Therefore, trisomic offspring heterozygous for the parental alleles indicate that crossing-over has occurred. It should be noted, however, that the converse of each case is not always true, as in meiosis I, the recombinant chromatids might not segregate into the functional meiotic product (Fig. 5). Similarly, chromatids with double crossovers between informative markers would not show evidence of recombination (Chakravarti and Slaugenhaupt, 1987).

For example, the results illustrated in Figure 6, derived from hybridization with the telomeric probe (D21S112), do not show evidence of recombination. Cytological markers indicate a meiosis I nondisjunction event occurred, and the trisomic child has three different alleles at the telomeric locus. The mother is heterozygous for alleles one and two, and the father carries alleles three and four. The trisomy 21 child inherited alleles 1 and 2 from her mother, and allele 4 from her father. Thus, the nondisjunction was maternal in origin, and since heterozygosity for the maternal alleles was retained following a meiosis I event, there is no indication that recombination occurred. Of the 9 total informative nondisjunctions found in this study, 6 were reduced to homozygosity, while 3 retained heterozygosity for the nondisjoined parental alleles at the q-terminal locus.

The telomeric marker was informative in 6 meiosis I events (Table 2). In four of these six, relevant markers were homozygous in the trisomic offspring, indicating a crossover had occurred. The remaining two retained heterozygosity even in this q-terminal region. Conversely, in three meiosis II events, one was found to be heterozygous ("CO") and two were homozygous ("NCO"). Hence, a total of five crossovers were detected in the nine informative nondisjunction events.

Table 2. Distribution of Recombination Among Informative Families With a DS Proband

	CO	NCO
MI	4	2
MII	1	2
TOTAL	5	4

Table 3. Double NOR Positive Parents*

Down syndrome	8/40 (20%)
Control	2/14 (14%)

*At least 10% of cells examined had a double NOR
$X^2 = 0.22$ (NS)

Double Nucleolar Organizing Regions (dNOR's)

No significant difference was detected between the frequency of double NOR's in 40 Down syndrome parent couples and 14 control couples ($X^2 = 0.22$; $p \gg 0.05$) (Table 3). Where peripheral blood samples from the Down syndrome child were available, dNOR positive parents were analyzed for the origin of nondisjunction. Two couples were found to be discordant for these two parameters (Table 4). In one of these couples, the father had the highest frequency of dNORs found in the study (60%), but nondisjunction took place in the mother.

The second case of discordance is illustrated by Family Q (Fig. 7) in which there were two children with Down syndrome. The father was dNOR positive in one of ten cells examined. However, as indicated by the highly informative telomeric DNA marker, only one nondisjunction was paternal, and the other was maternal in origin. An alternative explanation is that one parent is mosaic for a trisomic germ cell line.

Recombination was detected for one dNOR positive family, and indicates the utility of combining information from both cytological and molecular investigations. The mother in Family O (Fig. 8) was concordant for nondisjunction and dNOR's. The nondisjunction occurred in meiosis I, as indicated by chromosome heteromorphisms.

Table 4. Concordance of dNOR's and Nondisjunction

		dNOR Positive Parent	
		Mat	Pat
Origin of NDJ	Mat	3	2
	Pat	0	1

Family Q

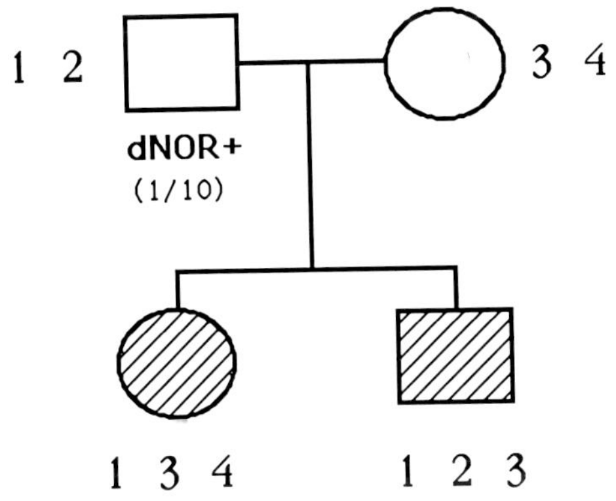

Figure 7. Family Q. Genotypic differences in trisomic offspring detected by the telomeric probe.

Maternal nondisjunction was demonstrated with results from both the D21S3 and telomeric probes. The mother was heterozygous at the telomeric locus but the child was homozygous for the maternal telomeric alleles, indicating a crossover had occurred between the cytological markers and the telomere in the nondisjoined chromosomes.

Maternal Age

The maternal age at birth of the Down syndrome child did not vary according to either parental origin of nondisjunction, or recombinant class. The cases were evenly distributed for mothers older and younger than 30 years of age, with an average of 30 (range 18-42). Only three mothers were over 35 years old. Similarly, when examined according to origin of nondisjunction, the ages were evenly distributed above and below 30 for all categories (maternal, paternal, and meiosis I and II events).

Family O

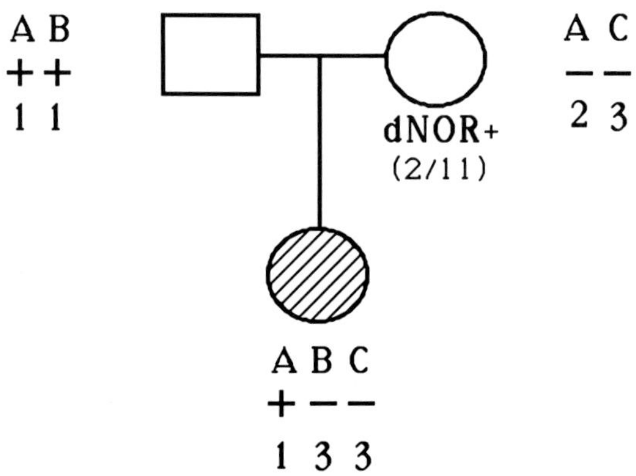

Figure 8. Family O. Pedigree indicating maternal meiosis I nondisjunction with recombination between the cytological markers and D21S112. Chromosomal heteromorphisms are designated by capital letters, and are specific within a family. The RFLP at D21S3 is designated as "+" or "-", and the telomeric marker is designated by Roman numerals. No linkage phase is implied by the order shown.

Using the Wilcoxon rank test, no significant differences were found between parental ages for maternal vs. paternal, or meiosis I vs. II nondisjunction events ($p>0.5$) (Table 5a). In fact, the mean ages are very close, and differ by only one year in each group. Similarly, for those families demonstrating maternal nondisjunction, the average maternal age for the recombinant group does not differ from the group of "non-recombinants" (Table 5b).

DISCUSSION

The results of this series of investigations clearly indicate the value of using an approach which combines molecular and cytological methods to study cytogenetic phenomena. The cytological markers employed increased the

Table 5a. Average Maternal Age by Origin of NDJ

Origin	N	$\bar{x}$	S.D.	
MMI	10	29.51	5.11	
All maternal	13	30.12	4.98	
All paternal	3	31.45	1.48	NS
Meiosis I	13	30.21	5.81	
Meiosis II	4	30.95	4.23	NS

Table 5b. Recombination and Age in Maternal Nondisjunction

Origins	N	$\bar{x}$	
Recombinants	5	29.7	
"Non-recombinants"	2	29.6	NS

number of informative families, especially with respect to the stage of meiosis in which nondisjunction occurred. The molecular markers, when informative, either support or pinpoint the parental origin of nondisjunction.

The availability of well-mapped molecular markers spanning the long arm of chromosome 21 enables the determination of recombination frequency for chromosomes undergoing nondisjunction. This significantly contributes to our understanding of the etiology of nondisjunction, by making possible the analysis of events in different meiotic stages, and with respect to age and sex of the parent. The presence of the dNOR variant could be analyzed for association with recombination only by using the combined cytological and molecular techniques.

Evidence for recombination in the nondisjoined chromosomes was observed in five of nine informative Down syndrome families. The finding of 4/6, or 67%, recombinants in chromosomes undergoing nondisjunction in meiosis I indicates a higher rate of recombination than reported previously. Warren et al (1987) found no evidence of recombination in nondisjoined chromosomes from nine meiosis I events. In our sample of families, further work is currently in progress to localize the chromosomal region of recombination in each family. However, at least one crossover occurred distal to the most telomeric locus used by Warren (D21S3/S23). Another study using a combination of cytological and molecular markers also reported recombination on the nondisjoined chromosomes -- the supernumerary chromosomes in two trisomy 21 offspring were determined to arise from maternal meiosis I nondisjunction, and each trisomic child appeared to show evidence of recombination (Hamers et al, 1987).

The dNOR results of our study do not support the previous findings of an increased frequency of dNOR's in parents of children with Down syndrome. Our results differ from the findings of Jackson-Cook in two main respects. First, we found 14% dNOR positive individuals among our control couples. Jackson-Cook et al (1985), in a study of 50 couples with a Down syndrome child, found 15 (30%) with a dNOR, as opposed to none in a random sample of 50 individuals. Additionally, we found two of six cases discordant for dNOR's and nondisjunction while Jackson-Cook found no discordance in 13 families. Two other studies also contradict Jackson-Cook's results. Spinner et al (1986) found no significant difference in the frequency of the dNOR variant between 24 couples with a trisomy 21 offspring and 19 control couples. Six of the individuals with a trisomy 21 offspring had a dNOR (25%) compared to 6 of the control couples (32%). In addition, Hassold et al (1987) have reported 23 couples with a fetus trisomic for an acrocentric chromosome and 55 couples with karyotypically normal or nonacrocentric aneuploid spontaneous abortions. There was no significant difference between the two groups as no dNOR's were found in either group. Our data confirm the findings of the latter two studies, as we found no significant difference between our control couples compared to those with a trisomy 21 offspring. Furthermore, our data resemble Spinner's (1986) in that we found dNOR's in both groups studied.

We have found no difference in frequencies for dNOR's which would indicate a role in the etiology of trisomy 21. Moreover, two dNOR positive parents of Down syndrome children were nonconcordant for the origin of nondisjunction. We also have documented that recombination may occur, even in concordant cases, and hence, if the dNOR has a causal effect, its mechanism of action must be other than through the prevention of proper synapsis of homologues.

The main conclusion that can be drawn from these findings is that mechanisms or events in meiosis leading to a "non-recombinant" class of individuals are not entirely, or even significantly, responsible for meiosis I errors involving chromosome no. 21. Events following recombination, and after dictyotene, may well prove most important in the etiology nondisjunction. Further analysis of recombination in families with Down syndrome individuals undoubtedly will contribute further to our understanding of the extent of heterogeneity in the etiology of nondisjunction.

ACKNOWLEDGEMENTS

We would like to thank those who made their probes available to us: Drs. D.M. Kurnit for p214UNC (D21S110); J.F. Gusella for pPW-228C, 236B, 231F, and 244D (D21S1,S11,S3 and S23, respectively); and Collaborative Research, Inc. for L427 (D21S112).

REFERENCES

Caspersson T, Lomakka G, Zech L (1971). The 24 fluorescence patterns of the human metaphase chromosomes -- distinguishing characters and variability. Hereditas 67:89-102.

Chakravarti A, Slaugenhaupt SA (1987). Methods of studying recombination on chromosomes that undergo nondisjunction. Genomics 1:35-42.

Donahue RP (1970). The relation of oocyte maturation to ovulation in mammals. In Biggers JD, Schuetz AW (eds): "Oogenesis", Baltimore: University Park Press, pp 413-438.

Donis-Keller H, Green P, Helms C, Cartinhour S, Weiffenbach B, Stephens K, Keith TP, Bowden DW, Smith DR, Lander ES, Botstein D, Akots G, Rediker KS, Gravius T, Brown VA, Rising MB, Parker C, Powers JA, Watt DE, Kauffman ER, Bricker A, Phipps P, Muller-Kahle H, Fulton TR, Ng S, Schumm JW, Braman JC, Knowlton RG, Barker DF, Crooks SM, Lincoln SE, Daly MJ, Abrahamson J (1987). A genetic linkage map of the human genome. Cell 51:319-337.

Feinberg AP, Vogelstein B (1983). A technique for radiolabeling DNA restriction endonuclease fragments to a high specific activity. Anal Biochem 132:266-267.

Gosden JR, Lawrie SS, Gosden CM (1981). Satellite DNA sequences in the human acrocentric chromosomes: information from translocations and heteromorphisms. Am J Hum Genet 33:243-251.

Gould SL, Martin-Deleon PA (1987). BrDU-Giemsa labeling studies of satellite association in parents of children with trisomy 21 or 13. Am J Med Genet 26:971-981.

Grell RF (1985). The meiotic process and aneuploidy. In Dellarco VL, Voytek PE, Hollaender A (eds): "Aneuploidy: etiology and mechanism", New York: Plenum Press, pp 317-331.

Hamers AJH, Vaes-Peeters GPM, Jongbloed RJE, Millington-Ward AM, Meijer H, de Die-Smulders CEM, Geraedts JPM (1987). On the origin of recurrent trisomy 21: determination using chromosomal and DNA polymorphisms. Clin Genet 32:409-413.

Hassold T, Jacobs PA, Pettay D (1987). Analysis of nucleolar organizing regions in parents of trisomic spontaneous abortions. Hum Genet 76:381-384.

Hassold T, Warburton D, Kline J, Stein Z (1984). The relationship of maternal age and trisomy among trisomic spontaneous abortions. Am J Hum Genet 36:1349-1356.

Hassold TJ, Jacobs PA (1984). Trisomy in man. Ann Rev Genet 18:69-97.

Henderson SA, Edwards RG (1968). Chiasma frequency and maternal age in mammals. Nature 218:22-28.

Howell WM, Black DA (1980). Controlled silver-staining of nucleolus organizer regions with a protective colloidal developer: a 1-step method. Experientia 36:1014-1015.

Hubbell HR (1985). Silver-staining as an indicator of active ribosomal genes. Stain Technology 60:285-294.

Hulten M (1974). Chiasma distribution at diakinesis in the normal human male. Hereditas 76:55-78.

Ikeuchi T (1984). Inhibitory effect of ethidium bormide on mitotic chromosome condensation and its application to high resolution chromosome banding. Cytogenet Cell Genet 38:56-61.

ISCN (1985). "An International System for Human Cytogenetic Nomenclature: Report of the standing

committee on human cytogenetic nomenclature". New York: Karger, Inc., p 11.
Jackson-Cook CK, Flannery DB, Corey LA, Nance WE, Brown JA (1985). Nucleolar organizer region variants as a risk factor for Down syndrome. Am J Hum Genet 37:1049-1061.
Jacobs PA, Mayer M (1981). The origin of human trisomy: study of heteromorphisms and satellite association. Ann Hum Genet 45:357-365.
Juberg R, Mowrey P (1983). Origin of nondisjunction in trisomy 21 syndrome: all studies compiled, parental age analysis and international comparisons. Am J Med Genet 16:111-116.
Kan YW, Dozy AM, Trecartin R, Todd D (1977). Identification of a nondeletion defect in α-thalassemia. New Eng J Med 297:1081-1084.
Kaplan JC, Carritt B (1987). Report of the committee on the genetic consitution of chromosomes 20, 21, and 22. Nineth International Workshop on Human Gene Mapping. Cytogenet Cell Genet 46:257-276.
Kazazian HH, Jr., Antonarkis SE, Wong C, Trusko SP, Stetten G, Oliver M, Potter MJ, Gusella JF, Watkins PC (1985). Ring chromosome 21: characterization of DNA sequences at sites of breakage and reunion. Ann NY Acad Sci 450:33-42.
Korenberg JR, Engels WR (1978). Base ratio, DNA content and quinacrine-brightness of human chromosomes. Proc Natl Acad Sci USA 75:3382-3386.
Kurnit DM, Stewart GD, Roy S, Schweduck J, Neve RL, Van Keuren M, Patterson D, Tantravahi U, Latt SA (1986). The centromeric region of chromosome 21: molecular and clinical studies on nondisjunction. Ped Res 617:735-7577.
Kurnit DM, Neve RL, Morton CC, Bruns GAP, Ma NSF, Cox DR, Klinger HP (1984). Recent evolution of DNA sequence homology in the pericentromeric regions of human acrocentric chromosomes. Cytogenet Cell Genet 38:99-105.
Lejeune J, Gauthier M, Turpin MR (1959). Les chromosomes humains en culture de tissus. C R Acad Sci, Paris 248:602-603.
Licznerski G, Lindsten J (1972). Brief reports: Trisomy 21 in man due to maternal non-disjunction during the first meiotic division. Hereditas 70:153-154.

Markovic VD, Worton RG, Berg JM (1978). Evidence for the inheritance of silver-stained nucleolus organizer regions. Hum Genet 41:181-187.

Melnyk JH, Persinger GW, Coss J, Murphy DE (1987). Down syndrome and NORs. 25th Annual Somatic Cell Genetics Conference. Santa Fe, New Mexico.

Mikkelsen M, Poulsen H, Grinsted J, Lange A (1980). Nondisjunction in trisomy 21: study of chromosomal heteromorphisms in 110 families. Ann Hum Genet 44:17-28.

Penrose LS (1933). The relative effects of paternal and maternal age in mongolism. J Genet 27:219-224.

Polani PE, Briggs JH, Ford CE, Clarke CM, Berg JM (1960). A mongol girl with 46 chromosomes. Lancet pp 721-724.

Rasmussen SW, Holm PB (1985). The synaptonemal complex, recombination nodules and chiasmata in human spermatocytes. Symp Soc Exp Biol 38:271-291.

Southern EM (1975). Detection of specific sequences among DNA fragments separated by gel electrophoresis. J Mol Biol 98:503-517.

Spinner NB, Eunpu DL, Schmickel RD, Zackai E, Bunin G, Emanuel BS (1986). The role of cytologic and molecular NOR variants in trisomy 21. Am J Hum Genet 39:A133.

Tanzi RE, Haines JL, Watkins PC, Stewart GD, Wallace MR, Hallewell R, Wong C, Wexler NS, Conneally PM, Gusella JF (1988). Genetic linkage map of human chromosome 21. Genomics 3:129-136.

Uchida IA (1973). Paternal origin of the extra chromosome in Down's syndrome. Lancet ii: 1257-1258.

Warren AC, Chakravarti A, Wong C, Slaugenhaupt SA, Halloran SL, Watkins PC, Metaxotou C, Antonarakis S (1987). Evidence of reduced recombination on the nondisjoined chromosomes 21 in Down syndrome. Science 237:652-654.

Warren AC, Slaugenhaupt SA, Lewis JG, Chakravarti A, Antonarakis SE (1989). A genetic linkage map of human chromosome 21. Genomics (In press).

Watkins PC, Tanzi RE, Gibbons KT, Tricoli JV, Landes G, Eddy R, Shows TB, Gusella JF (1985). Isolation of polymorphic DNA segments from human chromosome 21. Nucleic Acids Res 13:6075-6088.

Weisblum B (1973). Why centric regions of quinacrine-treated mouse chromosomes show diminshed fluorescence. Nature 246:150-151.

Molecular and Cytogenetic Studies of
Non-Disjunction, pages 101–113

STUDYING NONDISJUNCTION OF CHROMOSOME 21 WITH CYTOGENETIC MARKERS ON THE SHORT ARM AND DNA MARKERS ENCOMPASSING THE LONG ARM

Monika Buraczynska[*], Gordon D. Stewart[*], Stephanie Sherman[†], Viola Freeman[#], Meg Grantham[†], Irene Uchida[#], Terry Hassold[†] and David M. Kurnit[*¶].

[*]Howard Hughes Medical Institute and the Dept. of Pediatrics and [¶]Dept. of Human Genetics, University of Michigan, Ann Arbor, MI 48109. [†]Emory University, Dept. of Pediatrics, Atlanta, GA 30322. [#]McMaster University, Dept. of Pediatrics, Hamilton, Ontario, Canada L8S 3Z5.

INTRODUCTION

Despite the high rate of occurrence and obvious clinical importance of Down syndrome, surprisingly little is known about the mechanism of origin of the extra chromosome 21 or about factors that predispose to nondisjunction involving chromosome 21. For example, studies of chromosome markers have shown that trisomy 21 usually results from nondisjunction at maternal meiosis I (Hassold et al., 1984). However, we do not know if this results from an error in chromosome pairing and recombination, chromosome separation or a completely different mechanism; we do not know if the primary event involves the spindle or the chromosome itself; and aside from maternal age, we remain ignorant of factors which directly affect the frequency of trisomy 21.

The introduction of molecular techniques to the study of chromosome organization makes it possible to study many of these questions, and this approach is especially efficacious when the molecular information is combined with cytogenetic observations. DNA probes which identify restriction fragment length polymorphisms (RFLPs) on chromosome 21 will permit rapid accumulation of objective information on the parental origin of trisomy 21 (Davies, et al., 1984; Stewart, et al., 1985; Stewart et al., 1988a, 1988b). The availability of multiple loci on chromosome 21 will make it possible to confirm parental origin assignments and, as RFLPs span the long arm of chromosome 21, they can be used to detect crossovers at any point along the chromosome. These data can then be combined with information on the meiotic stage of origin (from studies of chromosome heteromorphisms or centromeric DNA markers) to assess the role of errors of recombination in nondisjunction involving chromosome 21.

We have recently initiated a study of cytogenetic and DNA polymorphisms in trisomy 21 conceptuses and their parents in order to: (1) Determine the parental and meiotic stage of origin on the extra chromosome 21. (2) Determine the effect of increasing maternal age on nondisjunction at meiosis I and II and evaluate the hypothesis that the maternal age effect may be due to age related factors other than nondisjunction. (3) Evaluate the role of abnormal levels (reduced or increased) of recombination in nondisjunction of chromosome 21. (4) Construct male and female centromere maps of chromosome 21. (5) Assess the importance of maternal and paternal risk factors in the etiology of trisomy 21. (6) Evaluate the possibility that the parental origin of the extra chromosome 21 affects the phenotype of the trisomic offspring.

In the present paper, we summarize our observations on issues 1 - 3, based on an initial series of 30 trisomy 21 conceptuses and their parents. We were able to specify the parental origin in 22 cases, with 20 being maternal and 2 paternal in origin. We observed crossing over in 8 of the maternally derived cases, demonstrating that failure to pair is not the only mechanism responsible for trisomy 21. However, the number of families and loci thus far studied remains small, and it is not yet possible to evaluate critically the role of asynapsis, or abnormally low or high levels of recombination, in nondisjunction involving chromosome 21.

MATERIAL AND METHODS

The study population: The study population consisted of 30 trisomy 21 conceptuses and their parents. Five cases involved spontaneously aborted trisomic fetuses which were ascertained in a cytogenetic study of spontaneous abortions (Hassold et al., 1980) while the remaining 25 cases were liveborn individuals with clinical features of Down syndrome.

Cytogenetic methodology: Routine cytogenetic studies were done on all cases and parents, using standard Q- or G-banding procedures. One spontaneous abortion (E5, see Table 1) was a double trisomy, 48,XX,+7,+21, but all other cases had either a 47,XY,+21 or 47,XX,+21 chromosome constitution. All parents had apparently normal chromosome constitutions.

Chromosome 21 heteromorphisms of the parents were compared with those from their trisomic offspring using Q-banding with dichloromethoxyacridine/spermine (Degau, 1978). In some cases, the slides had been silver stained for analysis of nucleolar organizing region (NOR) variants before being Q-banded, making it possible to examine NOR and Q-heteromorphisms simultaneously. In all cases, the slides were analyzed by at least two independent observers and, in the event of a disagreement, a conservative consensus was taken to be the final decision.

DNA studies: DNA from fetal tissue, cultured fetal cells, and peripheral blood samples was prepared as previously described (Hassold et al., 1985; Stewart et al., 1988a). DNA samples were digested with the appropriate restriction enzymes according to the manufacturer's specifications (New England Biolabs) and size-fractionated by electrophoresis on 1% agarose gels in TBE buffer. The samples were then transferred to Genatran filters (Plasco, Inc.) or to Hybond (Amersham Corp.) using previously described procedures (Neve et al., 1986) and, after prehybridization, hybridized to probes which had been labelled using the random primer method of Feinberg and Vogelstein (1983). More recently the Amersham hybridization kit has been used, as described by the manufacturers.

Nineteen probes detecting chromosome 21 restriction fragment length polymorphisms (RFLPs) were used in the study. Eight of these (D21S11, D21S13, D21S15, D21S19, D21S110, D21S112, D21S113, and CD18) were applied to filters from all, or most of the families, while the remaining eleven probes (SOD1, D21S1, D21S8, D21S16, D21S17, D21S23, D21S52, D21S53, D21S58, D21S118 and D21S141) were used only in family studies of the five trisomy 21 spontaneous abortions. Results on these five families have been reported in detail previously (Stewart et al., 1988a).

RESULTS

Origin of nondisjunction: Details of cytogenetic and molecular studies of nondisjunction in the 30 families are provided in Table 1 and the results summarized in Table 2. In most instances, the DNA studies consisted of analyses at five to eight different polymorphic loci, including two proximal loci located in band 21q11 (D21S110 and D21S13), one more distally located in 21q21 (D21S11), and four located in the most distal band 21q22.3 (D21S15, D21S19, CD18 and D21S112). Using these markers, we were able to specify the parental origin in 16 of the 30 cases.

TABLE 1 - Cytogenetic and Molecular observations on 30 Trisomy 21 conceptions and their parents

FAMILY NUMBER	CYTOGENETIC ANALYSIS		DNA ANALYSIS Locus:								ORIGIN OF EXTRA 21
			D21S110	D21S13	D21S11	D21S15	D21S19	D21S113		D21S112	
		Probe:	4U	pGSM21	236B	E8	B3	pMCT15	CD18	CRI4-427	
		Location:	cen-q11	cen-q11	q11.2-q21	q22.3	q22.3	q22.3	q22.3-qter	q22.3-qter	
		Enzyme:	Msp I	Taq I	Taq I	Msp I	Pst I	Msp I	Pst I	Rsa I	
T 12 *	Fa	AA	1,2	1,2	1,1	1,2		1,2		ab	
	Mo	AB	2,2	1,1	1,2	1,2		2,2		ac	Maternal I
	+21	AAB	2,2,2	1,1,1	1,1,2	1,1,2		2,2,2		acc	
T 19	Fa	AA	1,2	2,2	2,2			2,2	2,2		
	Mo	AB	2,2	1,2	2,2			1,2	2,2	-	Unknown
	+21	AAB	2,2,2	1,2,2	2,2,2			1,2,2	2,2,2		
T 41	Fa	AA	2,2	1,1	1,2			1,1	2,2		
	Mo	AA	2,2	1,2	1,1			1,2	1,2	-	Unknown
	+21	AAA	2,2,2	1,1,2	1,1,1			1,1,2	1,2,2		
T 42 *	Fa	AA	1,2,	1,1,	1,1,	1,2	1,2	1,2	1,2	-	
	Mo	AB	2,2,	1,2	2,2	2,2	1,2	2,2	1,2	ab	Maternal I
	+21	AAB	2,2,2	1,1,2	1,2,2	1,2,2	1,1,2	1,2,2	1,2,2	aac	
T 43 **	Fa	AA	2,2	1,2	1,2	1,2	1,2	1,1	1,1	ab	
	Mo	AB	2,2,	1,2	1,2	1,1	1,2	1,2	1,2	ac	Maternal I
	+21	AAB	2,2,2	1,1,2	1,2,2	1,1,1	1,1,2	1,2,2	1,1,2	aac	
T 44 *	Fa	AB	2,2,	2,2,	2,2	1,2	1,2	1,1	2,2	-	
	Mo	AB	2,2	1,1	1,2	1,2	1,2	1,2	1,2	ab	Maternal
	+21	AAB	2,2,2	1,1,2	1,2,2	1,1,2	1,1,2	1,1,2	1,2,2	aac	
T 45	Fa	AB	1,2	1,2	1,2	2,2		1,1		ab	
	Mo	CD	1,1	-	-	1,2		1,1		ac	Maternal I
	+21	ACD	1,1,1	1,1,1	1,1,2	1,2,2		1,1,1		aac	
T 47 **	Fa	AB	2,2	1,1		1,1	1,2	2,2	1,2		
	Mo	AC	1,1	1,2		1,2	1,2	1,1	1,2	-	Maternal I
	+21	ABC	1,1,2	1,2,2		1,1,2	1,1,2	1,1,2	1,2,2		
T 49 *	Fa	AB	2,2	1,2		2,2		1,2	1,2	-	
	Mo	CD	2,2	1,2		1,1		2,2	1,2	ab	Maternal I
	+21	ACD	2,2,2	1,2,2		1,1,2		1,2,2	1,1,2	aac	
T 50	Fa	AA	2,2	1,1		2,2		1,1	2,2	ab	
	Mo	AA	1,2	-		1,2		1,2	1,2	ac	Unknown
	+21	AAA	1,2,2	1,1,1		1,2,2		1,1,2	1,2,2	aaa	

TABLE 1 - (con't)

FAMILY NUMBER	CYTOGENETIC ANALYSIS		Locus: / Probe: / Location: / Enzyme:	D21S110 / 4[illegible]J / cen-q11 / Msp I	D21S13 / pGSM21 / cen-q11 / Taq I	D21S11 / 236B / q11.2-q21 / Taq I	D21S15 / E8 / q22.3 / Msp I	D21S19 / B3 / q22.3 / Pst I	D21S113 / pMCT15 / q22.3 / Msp I	CD18 / q22.3-qter / Pst I	D21S112 / CRI4-427 / q22.3-qter / Rsa I	ORIGIN OF EXTRA 21
				DNA ANALYSIS								
	Fa	AB		2,2	1,1		2,2		2,2	1,2	ab	
T 51	Mo	AC		1,2	1,1		1,1		1,1	2,2	ac	Maternal I
	+21	ABC		1,2,2	1,1,1		1,1,2		1,1,2	1,2,2	abc	
	Fa	AB		2,2	1,1		1,2		1,1	1,2	ab	
T 52	Mo	AC		1,1	1,1		1,2		1,2	2,2	-	Maternal I
	+21	AAC		1,1,2	1,1,1		1,2,2		1,1,2	2,2,2	aaa	
	Fa	AB		1,1	1,1		1,2		1,2	2,2	ab	
T 53	Mo	AC		2,2	1,2		2,2		1,2	2,2	ac	Paternal II
	+21	AAC		1,1,2	1,1,2		2,2,2		1,1,2	2,2,2	aac	
	Fa	AA		1,2	1,2					1,2	ab	
T 54	Mo	AA		2,2	2,2					2,2	ac	Unknown
	+21	AAA		2,2,2	2,2,2					1,2,2	abc	
	Fa	AB		1,2		1,2	2,2		1,2	1,1	ab	
T 55	Mo	AC		2,2		1,2	2,2		1,2	1,2	ac	Unknown
	+21	ABC		2,2,2		1,2,2	2,2,2		1,2,2	1,1,2	aac	
	Fa	AB		2,2	1,1	2,2	1,2		1,2	2,2	ab	
T 56	Mo	CD		1,1	1,2	2,2	2,2		1,2	1,2	ac	Maternal I
	+21	ACD		1,1,2	1,1,2	2,2,2	2,2,2		1,2,2	1,2,2	aac	
	Fa	AB		2,2	1,1	1,1	1,1		1,2	1,1	ab	
T 57	Mo	CC		1,1	1,2	1,2	2,2		1,2	1,2	ac	Maternal
	+21	ACC		1,1,2	1,1,2	1,1,2	1,2,2		1,1,2	1,1,2	aac	
	Fa	AA		1,2	1,2	1,2	1,2		1,2	1,1	ab	
T 58	Mo	AA		2,2	1,1	1,2	1,2		1,1	1,2	ac	Unknown
	+21	AAA		2,2,2	1,1,1	1,1,2	1,2,2		1,1,1	1,1,2	abc	
	Fa	AA		1,2	1,1	1,1	2,2		1,2			
T 59	Mo	BC		1,2	1,2	1,2	2,2		1,2		-	Maternal I
	+21	ABC		1,2,2	1,1,2	1,1,2	2,2,2		1,1,2			
	Fa	AB		1,2	1,2	2,2	2,2		2,2	2,2	ab	
T 60	Mo	AC		1,2	1,1	2,2	1,2		1,1	1,2	ac	Maternal I
	+21	ABC		1,2,2	1,1,1	2,2,2	1,2,2		1,1,2	1,2,2	aac	

TABLE 1 - (con't)

FAMILY NUMBER	CYTOGENETIC ANALYSIS		DNA ANALYSIS								ORIGIN OF EXTRA 21
		Locus:	D21S110	D21S13	D21S11	D21S15	D21S19	D21S113		D21S112	
		Probe:	4U	pGSM21	236B	E8	B3	pMCT15	CD18	CRI4-427	
		Location:	cen-q11	cen-q11	q11.2-q21	q22.3	q22.3	q22.3	q22.3-qter	q22.3-qter	
		Enzyme	Msp I	Taq I	Taq I	Msp I	Pst I	Msp I	Pst I	Rsa I	
	Fa	AB	1,2	1,1	2,2	2,2		1,2			
T 61	Mo	CD	2,2	1,1	2,2	1,2		1,2		-	Maternal I
	+21	ACD	2,2,2	1,1,1	2,2,2	1,2,2		1,1,2			
	Fa	AA	2,2	1,2	1,2	1,2		1,2		ab	
T 62	Mo	BC	1,1	2,2	2,2	1,2		1,1		cc	Maternal I
	+21	ABC	1,1,2	2,2,2	2,2,2	1,1,2		1,1,1		acc	
	Fa	AA	1,2	1,2	2,2	2,2		1,1		ab	
T 64	Mo	BC	1,2	2,2	2,2	1,2		1,2		ac	Maternal I
	+21	ABC	1,2,2	2,2,2	2,2,2	1,2,2		1,1,2		aac	
	Fa	AB	2,2	1,2	1,2			1,1	-	ab	
T 70	Mo	AC	2,2	1,2	2,2			1,1	1,2	ac	Unknown
	+21	ABC	2,2,2	1,1,2	1,2,2			1,1,1	1,1,2	aac	
	Fa	AB	2,2	1,1	1,2			1,1	1,2		
T 71	Mo	AC	2,2	1,2	2,2			1,2	2,2	-	Unknown
	+21	ABC	2,2,2	1,1,2	2,2,2			1,1,2	2,2,2		
	Fa	AA	1,2	1,1	1,1	2,2	2,2				
E 1	Mo	AB	2,2	1,1	1,2	2,2	2,2			-	Paternal#
	+21	AAB	2,2,2	1,1,1	1,1,2	2,2,2	2,2,2				
	Fa	AA	-	-	1,1	2,2	2,2				
E 2	Mo	BC	2,2	-	1,2	-	2,2			-	Maternal I
	+21	ABC	1,2,2	1,1,1	1,1,2	2,2,2	2,2,2				
	Fa	AB	1,1	1,1	1,1	1,1	2,2				
E 3	Mo	CD	1,1	1,1	1,2	1,2	2,2			-	Maternal I
	+21	ACD	1,1,1	1,1,1	1,1,2	1,1,2	2,2,2				
	Fa	AB	1,2	2,2	1,2	1,2	1,2				
E 4 *	Mo	AC	1,2	1,1	1,2	2,2	1,2			-	Maternal I#
	+21	ABC	1,1,2	1,1,2	1,1,2	1,2,2	1,2,2				
	Fa	AB	2,2	1,2	1,2	1,2	2,2				
E 5 *	Mo	AA	2,2	1,2	1,2	2,2	1,2			-	Maternal#
	+21	AAB	2,2,2	2,2,2	1,1,2	1,2,2	1,2,2				

Parental origin assignments made on basis of results at DNA markers not shown in table (in the case of E4 a crossover demonstrated using such a marker).

* Single (or odd number) crossover identified

** Double (or even number) crossover identified

\- Uninformative

The cytogenetic studies resulted in parental origin determinations in ten cases, of which five were confirmations of the DNA studies. In no instance was there a discrepancy in parental origin assignment between the cytogenetic and molecular analyses.

By combining the results of the chromosome heteromorphism studies and the DNA marker studies, we were able to determine the parent of origin in 22 of the 30 families, with 20 being maternal and 2 paternal in origin. Furthermore, in 18 of these cases, we were also able to specify the meiotic stage of origin; 17 were due to a maternal meiosis I error and 1 to a paternal meiosis II error, with 3 other cases being due to an error at maternal meiosis I or II and 1 case to an error at paternal meiosis I or II.

TABLE 2. Summary of cytogenetic and molecular studies of nondisjunction in trisomy 21

Parent/meiotic state of origin*	No. of Cases	Mean maternal age (±SD)
Maternal I	17	30.6 ± 5.7
II	-	-
I or II	3	27.3 ± 2.1
Paternal I	-	-
II	1	28
I or II	1	38
Unknown+	8	26.3 ± 4.2

* In 12 cases the parental origin assignment was based solely on DNA analysis, in 6 cases solely on cytogenetic analysis, and in 4 cases on the basis of both DNA and cytogenetic analysis. In no instance was there a discrepancy between the DNA analysis and the final cytogenetic analysis.

+ In 4 of the 8 unknown cases, cytogenetic analysis ruled out at least one of the four meiotic stages.

Maternal age and nondisjunction: Information on maternal age is summarized by parent and meiotic stage of origin (Table 2). There were no significant mean differences between the various categories; however, it is clear that there are, as yet, too few data to draw meaningful comparisons.

Recombination and nondisjunction: All cases of known parental origin were evaluated for presence or absence of recombination between the two nondisjoined chromosomes. Trisomic conceptuses were scored as being nonreduced (heterozygous) or reduced (homozygous) for RFLPs known to be heterozygous in the parent of origin. In trisomy of meiosis I origin, reduction at a DNA locus indicates recombination between the centromere and the locus; in meiosis II trisomy, nonreduction of the DNA locus indicates recombination has occurred (Shahar and Morton, 1986).

In 18 cases, both the parent and meiotic stage of origin were known, making it possible to study recombination between the centromere and DNA markers as well as recombination between different RFLPs. However, in five instances in which the parent, but not the meiotic stage, was known we could only evaluate recombination between the RFLPs.

Recombinants were observed in six of the seventeen families of known maternal meiosis I origin and in two additional families of maternal meiosis I or II origin (Table 1). Furthermore, in two of these families, T43 and T47, we observed double crossover events. In Family T43, CD18 and D21S113 are fully informative; CD18 is nonreduced and D21S113 is reduced in the trisomic child. The likeliest order of the markers is centromere--D21S113--CD18 (Stewart, unpublished data), thus two recombinants are necessary to explain the observations; one between the centromere and D21S113 (reduced) and one between D21S113 and CD18 (nonreduced).

In family T47, D21S13 and D21S15 are fully informative and the order of markers relative to the centromere is centromere--D21S13--D21S15. Therefore, recombination must have occurred between the centromere and D21S13 and a second crossover between D21S13 and D21S15; thus the reduced contribution from the mother at D21S13 and the nonreduced contribution at D21S15.

We were unable to detect recombinants in any of the remaining 14 trisomies of known parental origin. However, in the majority of these cases there were no informative loci or information was available at only one RFLP.

DISCUSSION

Over the past 15 years, studies of chromosome heteromorphisms have been extensively used to study the origin of trisomy 21, and information from this source is available on over 1,000 Down syndrome families (Hassold et al., 1984). From these studies it is clear that trisomy 21 is more often maternal than paternal in origin, and most frequently the result of nondisjunction at maternal meiosis I. These observations were not entirely unexpected, since the association between maternal age and Down syndrome was first described over 50 years ago (Penrose, 1933). However, at least two of the implications of the cytogenetic studies of trisomy 21 were surprising: First, the observed level of paternal nondisjunction was approximately 20-25%, much higher than had been anticipated and, second, the proportion of maternally derived cases was approximately the same for younger as older women. This latter observation led, in part, to the "relaxed selection" hypothesis, which suggests that the maternal age effect in trisomy might be due to an age-related decrease in the ability to abort chromosomally abnormal conceptions (Ayme and Lippman-Hand, 1982) rather than age-related increases in the production of abnormal oocytes (i.e., an "older egg" model).

The validity of these implications depends on the reliability of chromosome heteromorphism analysis, and several investigators have suggested that imprecision in cytogenetic analyses might lead to spurious conclusions (e.g., Hassold, 1985; Carothers, 1987). Therefore, results of DNA studies of chromosome 21 nondisjunction have been awaited with interest, since RFLP analysis should be more objective than the cytogenetic tests, and since the availability of over 50 RFLPs on chromosome 21 makes it possible to confirm results at several loci.

In the present study, we have used this approach coupled with cytogenetic analyses to study nondisjunction in an initial series of 30 trisomy 21 conceptuses. We determined the parental origin in 22 cases, 16 on the basis of DNA polymorphisms. As 2 of these cases were paternally derived, it is clear that paternal nondisjunction contributes to trisomy 21. Furthermore, this observation has been corroborated in several other laboratories. In Table 3, we have summarized results from six recent studies on nondisjunction in trisomy 21, listing only those results which were based on DNA markers or DNA and cytogenetic markers. The frequency of paternal nondisjunction varies considerably among the studies, from a low of 10% to a high of 39%. However, it is likely that this is due to the small number of cases studied, and it now seems reasonable to conclude that the level of paternal nondisjunction is at least 15%, and may be as high as the 20%-25% value originally observed in the cytogenetic studies.

TABLE 3. Results of analyses of chromosome 21 nondisjunction based solely on DNA markers.

	Origin of nondisjunction		
Study	Paternal %	Maternal %	Unknown
Stewart et al (1985)	1 (33%)	2 (67%)	7
Warren et al (1987) and personal communication	6 (10%)	55 (90%)	0
Rudd et al (1988)	2 (22%)	7 (78%)	11
Bricarelli et al (1988)	11 (39%)	17 (61%)	9
Galt et al (in press)	3 (25%)	9 (75%)	21
Present study	2 (12%)	16 (88%)	12
TOTAL	25 (19%)	106 (81%)	60

The relationship between maternal age and parental origin of nondisjunction is not nearly as clear. In Table 4, we have summarized data from three studies in which maternal age was available for trisomy of known and unknown parental origin. The highest mean maternal age is associated with maternally derived trisomy, consistent with the idea that frequency of nondisjunction increases with advancing maternal age. However, neither the mean maternal ages nor the distribution of ages is significantly different among the three categories and, somewhat surprisingly, the mean maternal age is the same for the unknown group as for paternal trisomy. Therefore, it is not yet possible to discriminate, on the basis of these limited data, between the "relaxed selection" and the "older egg" models of the maternal age--trisomy relationship.

TABLE 4. Relationship between maternal age and parental origin of nondisjunction in cases of trisomy 21 studied with DNA markers*

Parental origin of trisomy	Maternal Age Distribution 20-24	25-29	30-34	35-39	≥40	Mean Maternal age ± S.D.
Paternal	1	4	1	1	0	29.3 ± 5.3
Maternal	5	12	10	6	9	31.9 ± 6.8
Unknown	7	11	8	1	3	29.3 ± 6.0

* Data from present study, Rudd et al (1988), and S. Antonarakis (personal communication).

In addition to the above studies of parental origin and parental age and origin, the availability of DNA markers has made it possible to evaluate the relationship between abnormal levels of recombination and nondisjunction. Using this approach, Warren et al (1987) and Roulston et al (1988) have recently suggested that absent or reduced levels of recombination may be important in nondisjunction involving chromosome 21. In the present study, we have already observed recombinants in 8 of 20 trisomies of maternal origin, even though the number of informative loci per case does not exceed 3, and is typically 1-2. These results are somewhat at variance with those of Warren et al (1987), who observed virtually no recombinants in their study of 34 Down syndrome families. However, both in their study and in ours, the number of families and loci examined is still quite small, and it is not yet possible to critically evaluate the role of asynapsis or abnormally low levels of recombination to nondisjunction involving chromosome 21. Additionally, the markers used in their study did not span the length of the long arm and had we used the same markers as they, we would have found fewer crossovers.

Furthermore, while considerable attention has focused on the possible link between reduced recombination and nondisjunction, it is equally important to consider the possibility that increased recombination resulting in "chromosome entanglement" may lead to trisomy, an idea first formulated by Bridges in 1916. In studies of Drosophila, increased frequency of recombination around the centromere has been associated with nondisjunction (Merriam and

Frost, 1964). Similarly in the present study, we have already observed one recombinant (in four informative meioses) between 21p and D21S13, a marker located very near the centromere. Additionally, in a study of trisomy 13, we detected a crossover between the centromere and the most proximal locus studied in 3 of 5 cases of maternal meiosis I origin (Hassold et al., 1987; Hassold, unpublished observations). These observations reenforce the need to examine carefully the pericentromeric region in trisomy 21 and, as no centromeric DNA markers on chromosome 21p are yet available, to continue to use cytogenetic markers to evaluate the meiotic stage of origin. With the study of ever increasing numbers of families, the data available will become more reliable and allow a more accurate picture of nondisjunction and recombination to emerge.

ACKNOWLEDGEMENTS

This research was supported by NIH grants R01 HD21341, R01 25509, GM 36295, and March of Dimes 6-501. DMK, MB and GDS are supported by the Howard Hughes Medical Institute.

REFERENCES

Ayme S, and Lippman-Hand A. (1982). Maternal-age effect in aneuploidy: does altered embryonic selection play a role? Am J Hum Genet 39:558-565.

Bridges CB (1916). Nondisjunction as proof of the chromosome theory of heredity. Genetics 1:1-52; 107-163.

Carothers A (1987). Down syndrome and maternal age: the effect of erroneous assignment of parental origin. Am J Hum Genet 40:147-150.

Davies KE, Harper K, Bronthron D, Krumlauf R, Polkey A, Pembry ME, Williamson R (1984). Use of a chromosome 21 cloned DNA probe for the analysis of nondisjunction in Downs syndrome. Hum Genet 66:54-56.

Degau JH, van de Sande JH, and Lin CC (1978). $(CMA)_2S$: a double intercalating analogue of quinacrine for chromosome banding. Am J Hum Genet 30:78A.

Feinberg AP, and Vogelstein B (1983). A technique for radiolabeling restriction fragments to high specific activity. Anal Biochem 132:6-13.

Hassold T, Chiu D, Yamane J (1984) Parental origin of autosomal trisomies. Ann Hum Genet 48:129-144.

Hassold T, Jacobs PA, Leppert M, Sheldon M (1987). Cytogenetic and molecular studies of trisomy 13. J Med Genet 24:725-732.

Hassold T, Chen N, Funkhouser J, Jooss T, Manuel B, Matsuura J, and Matsuyama A (1980). A cytogenetic study of 1,000 spontaneous abortions. Ann Hum Genet 44:151-178.

Hassold T, Kumlin E, Leppert M, Takaesu N (1985). Determination of the parental origin of sex chromosome monosomy using restriction fragment length polymorphisms. Am J Hum Genet 37: 956-972.

Merriam JR, Frost JN. (1964). Exchange and nondisjunction of the X-chromosomes in female Drosophila melanogaster. Genetics 49:109-122.

Neve RL, Stewart GD, Newcomb P, Van Keuren ML, Patterson D, Drabkin HA, Kurnit DM (1986). Human chromosome 21-encoded cDNA clones. Gene 49:361-369.

Penrose LS (1933). The relative effect of paternal age and maternal age in mongolism. J Genet 27:219-224.

Rudd NL, Dimnik LS, Greentree C, Mendes-Crabb K, Hoar DI (1988). The use of DNA probes to establish parental origin in Down syndrome. Hum Genet 78:175-178.

Roulston D, Cohen MM, Schwartz S (1988). Cytological and molecular family studies of double NORs and recombination in trisomy 21. Am J Hum Genet 43:A119, Suppl.

Shahar S, Morton NE (1986). Origin of teratomas and twins. Hum Genet 74:215-218.

Stewart GD, Harris P, Galt J, Ferguson-Smith MA (1985). Cloned DNA probes regionally mapped to human chromosome 21 and their use in determining the origin of nondisjunction. Nucl Acids Res 13:4125-4132.

Stewart GD, Hassold TJ, Berg A, Watkins P, Tanzi R, Kurnit DM (1988a). Down syndrome: studying nondisjunction and meiotic recombination using cytogenetic and molecular polymorphisms that span chromosome 21. Am J Hum Genet 42:227-236.

Stewart GD, Hassold TJ, Kurnit DM (1988b). Trisomy 21 molecular and cytogenetic studies of nondisjunction in Harris H , Hirschorn K (eds): "Advances in Human Genetics", New York: Plenum Publishing Corp., pp 99-140.

Warren AC, Chakravarti A, Wong C, Slaugenhaupt SA, Halloran SL, Watkins PC, Metaxotou C, Antonarakis SE (1987). Evidence for reduced recombination on the nondisjoined chromosome 21 in Down syndrome. Science 237:652-654.

Molecular and Cytogenetic Studies of Non-Disjunction, pages 115–134

ANALYSIS OF NON-DISJUNCTION IN HUMAN TRISOMIC SPONTANEOUS ABORTIONS

Terry J. Hassold and Norma Takaesu

Department of Pediatrics, Division of Medical Genetics, Emory University Medical School, Atlanta, Georgia 30322

INTRODUCTION

Autosomal trisomy is the most common type of chromosome abnormality in humans, occurring in at least 3 - 4% of all clinically recognized pregnancies. A small proportion survive to term, about 0.3% of newborns having an additional chromosome 13, 18, or 21, but the vast majority of autosomal trisomies are eliminated prenatally and account for 25% of all clinically recognized spontaneous abortions.

This high level of *in utero* death makes spontaneous abortions a particularly useful population for studying the origin and etiology of trisomy. Therefore, for the past several years, we have been conducting cytogenetic studies of spontaneous abortions to identify trisomies, and subsequently using cytogenetic and/or DNA polymorphisms to determine the parent and meiotic stage of origin of the additional chromosome. In initial studies based solely on cytogenetic techniques (Hassold et al., 1984), we reported that errors in oogenesis were the most likely source of trisomy, regardless of chromosome involved, but that the level of paternal non-disjunction varied among chromosomes. Specifically, we observed a high level of paternal non-disjunction for trisomy 21, and suggested that this might be due to mechanisms of paternal non-disjunction wholly, or largely restricted to chromosome 21.

More recently, we have begun to use restriction fragment length polymorphisms (RFLPs) to study the origin of trisomy, and have recently presented preliminary results on trisomies 13 and 21 (Hassold et al., 1988; Stewart et al., 1988). RFLP analysis is particularly useful for this type of study, since the availability of multiple, chromosome specific polymorphisms makes the determination of the parental origin of trisomy virtually certain. Furthermore, by combining chromosome heteromorphism studies with analyses of DNA markers, it is now possible to address questions which were previously intractable. For example, studies of other organisms, including yeast and Drosophila, have demonstrated an association between abnormal levels of meiotic recombination and non-disjunction (e.g., Surosky and Tye, 1988; Merriam and Frost, 1964). The analysis of RFLPs in trisomic individuals and their parents provides a mechanism for studying this question in humans and preliminary results using this approach suggest that there may be an association between abnormal pairing or recombination and non-disjunction for chromosome 21 (Warren et al., 1987). Specifically, Warren et al. (1987) studied 34 Down syndrome families, and found little evidence that crossing-over had occurred between the two non-disjoined chromosomes, suggesting that failure to pair was responsible for many of the trisomies.

In the present report, we summarize our observations on 310 spontaneously aborted autosomal trisomies, including 36 which were studied with both cytogenetic and DNA markers. Our results confirm previous observations of a preponderance of maternally derived trisomies, regardless of the specific chromosome, and suggest an unusually high level of paternal non-disjunction for trisomy 21. However, we observed at least one cross-over in most trisomies for which information was available, suggesting that absence of recombination is relatively unimportant in the genesis of human autosomal trisomy.

MATERIALS AND METHODS

The Study Population

From 1976-1985, we conducted a cytogenetic study of all spontaneous abortions occurring at a single hospital in Honolulu. A total of 2922 fetuses were karyotyped, with the frequency and distribution of the different chromosome constitutions recently being presented by Jacobs et al. (1988). 783 single trisomies were identified and, of these, 310 with an additional chromosome 3, 4, 9, 13, 14, 15, 16, 21, or 22 were further studied to determine the parent and meiotic stage of origin of trisomy. Cytogenetic and molecular studies involving a subset of these fetuses and their parents have been presented previously (Hassold et al., 1984a; Hassold et al., 1988; Stewart et al., 1988). The present report summarizes our observations on the entire series.

Cytogenetic Studies

Chromosome heteromorphisms of trisomic fetuses and their parents were studied using Q-banding and/or silver staining techniques for analysis of trisomies 3, 4, 13, 14, 15, 21, and 22 and with fluorescent or non-fluorescent C-banding techniques for analysis of trisomies 9 and 16. All observations were made directly on the microscope by at least two independent observers. The results were compared and in the event of a disagreement, the most conservative decision consistent with each set of observations was taken as the final one.

Molecular Studies

DNA was extracted from tissue or blood samples, and processed for hybridization to DNA probes using previously described techniques (Hassold et al., 1988). A total of 31 probes detecting restriction fragment length polymorphisms were used in the study, consisting of seven for trisomy 13 (D13S1, D13S2, D13S3, D13S4, D13S5, D13S7, D13S10), nine for trisomy 16 (HP, CTRB, APRT, HB, LC9, D16S35, D16S37, D16S83, D16S84), eleven for trisomy 21 (SOD1, D21S1, D21S8, D21S16, D21S17, D21S23, D21S52, D21S53, D21S58, D21S118, D21S141), and four for trisomy 22 (IGLC, D22S1, D22S9, and D22S10). Information on the alleles detected by these probes and the physical locations of the loci defined by the probes is available elsewhere (Willard et al., 1987; Stewart et al., 1988).

Analysis of Recombination between the Non-disjoined Chromosomes.

For trisomies in which the parent of origin was known, we evaluated the frequency of crossing-over between the two non-disjoined chromosomes by: 1) identifying loci for which the parent of origin of trisomy was heterozygous 2) studying these loci in the trisomic offspring to determine if heterozygosity had been maintained, or had been reduced to homozygosity and 3) assessing the frequency of heterozygosity/homozygosity at different loci in the trisomic offspring. For example, if the parent of origin were heterozygous at two loci and the trisomic offspring heterozygous at one and homozygous at the other, a single (or odd number of) cross-over(s) must have occurred between these loci.

In trisomies in which the parent and meiotic stage of origin had been determined, we used this approach to evaluate crossing-over between the centromere and DNA markers and between successive DNA markers; in trisomies in which the parent, but not the meiotic stage, of origin was known, we were only able to assess crossing-over between DNA markers. For the purpose of the present analysis, we considered trisomies as being positive for recombination if one or more cross-overs were detected and negative for recombination if at least four informative loci were identified with no evidence for crossing-over between the loci. All other trisomies were considered to be uninformative.

RESULTS

Cytogenetic Studies

A summary of the cytogenetic studies of the parent and meiotic stage of origin is presented in Table 1. The probability of specifying the parent of origin varied considerably among trisomies, from a low of 0% (0/9) for trisomy 9 to a high of 65.6% (19/29) for trisomy 13. In total, we studied 310 single autosomal trisomies, and in 92 (29.7%) of these we were able to determine the parental source of the extra chromosome.

Table 1. Summary of results from cytogenetic studies of parental origin of 310 single autosomal trisomies.

		Meiotic Origin of Trisomy						
	No. of	Paternal			Maternal			
Trisomy	cases	I	II	I or II	I	II	I or II	%Paternal
3	6	-	-	-	1	-	-	0.0
4	15	-	-	-	1	-	-	0.0
9	9	-	-	-	-	-	-	-
13	29	2	1	-	9	4	3	15.8
14	12	-	-	-	2	1	-	0.0
15	22	-	-	-	5	-	2	0.0
16	115	1	-	-	17	2	4	4.2
21	38	-	-	2	6	-	3	18.2
22	64	1	-	1	19	-	5	7.7
TOTAL	310	4	1	3	60	7	17	
		8			84			8.7

Maternal errors were much more frequent than were paternal errors, and accounted for over 90% of all trisomies. The single most common mechanism of origin was non-disjunction at maternal meiosis I, which was observed in 60 cases. By comparison, we identified only seven errors at maternal meiosis II, four at paternal meiosis I, and only one case was attributable to an error at paternal meiosis II.

The high rate of maternal non-disjunction was observed for all trisomies and in no trisomy were there more paternal than maternal errors. However, the level of paternal non-disjunction varied considerably among trisomies. Specifically, the frequency of paternal non-disjunction was 15.8% (3/19) for trisomy 13 and 18.2% (2/11) for trisomy 21, but was only 4.8% (3/62) for the remaining trisomies.

Table 2 summarizes information on maternal age for trisomies of known parental origin. There were no significant differences in mean maternal age between these cases and trisomies in which we were unable to determine the parental origin, nor were there significant differences between these cases and trisomies identified in the general survey of spontaneous abortions (Hassold et al., 1984b). The mean maternal age was virtually identical for cases of paternal and maternal origin, with respective values of 31.5 $\pm$ 6.0 years and 31.9 $\pm$ 5.4 years. Both of these values were significantly higher than the mean maternal age of 26.1 $\pm$ 5.4 years observed for 41,705 newborn infants delivered at the study hospital over a comparable period (for paternal non-disjunction, $t = 2.55$, $p<0.05$; for maternal non-disjunction, $t = 8.92$, $p<0.001$).

Cytogenetic and Molecular Studies of Trisomies 13, 16, 21, and 22

In 36 cases of trisomies 13, 16, 21, or 22, DNA was obtained from the trisomic conception and parents, making it possible to study both cytogenetic and DNA markers. The results of the studies of the individual cases are summarized in Table 3, and are described below for each of the four different trisomies.

Table 2. Mean maternal ages in trisomies in which the parent and meiotic stage of origin was determined using cytogenetic markers.

Trisomy	PATERNAL I	PATERNAL II	MATERNAL I	MATERNAL II
3	-	-	26 (1)*	-
4	-	-	19 (1)	-
13	26.5 ± 12.0 (2)	32 (1)	32.4 ± 6.1 (9)	36.5 ± 7.9 (4)
14	-	-	31.0 ± 7.1 (2)	40 (1)
15	-	-	31.8 ± 10.2 (5)	-
16	38 (1)	-	31.1 ± 4.6 (17)	36.5 ± 7.8 (2)
21	-	-	32.5 ± 4.0 (6)	-
22	35 (1)	-	33.3 ± 4.5 (19)	-
TOTAL	31.5 ± 9.1 (4)	32 (1)	31.9 ± 5.4 (60)	37.0 ± 6.6 (7)
	31.5 ± 6.0 (8)+		31.9 ± 6.0 (84)#	

* Number of cases indicated by ( ).

+ Includes two cases of trisomy 21 and one case of trisomy 22 in which the origin was paternal meiosis I or II.

\# Includes three cases of trisomy 13, two of trisomy 15, four of trisomy 16, three of trisomy 21, and five of trisomy 22 in which the origin was maternal meiosis I or II.

Table 3. Summary of cytogenetic and molecular studies of 36 spontaneous abortions with trisomies 13, 16, 21, or 22.

I.D. No.	Chromosome Constitution	Parental Ages Fa	Parental Ages Mo	Origin of Trisomy* Cytogenetics	Origin of Trisomy* DNA	Evidence for Recombination+
K3177	47,XX,+13	36	28	Mat I		Yes
K3283	47,XX,+13	35	39	Mat I		
K3325	47,XY,+13	36	38	Mat I		Yes
K3392	47,XY,+13	29	27	Mat II	Mat	Yes
K3362	46,X,+13	36	30	Mat II or Pat I	Pat	
K2766	47,XY,+13	43	35	Pat I		Yes
K2912	47,XX,+16	30	25		Pat	Yes
K3021	45,X/46,X,+16	34	34		Mat	
K3029	47,XX,+16	42	30		Mat	No#
K3056	47,XX,+16	37	35	Mat I, II or Pat II	Mat	No#
K3060	47,XX,+16	36	38	Pat I		
K3066	47,XY,+16	30	29	Mat I or Pat I		

Table 3. (cont'd.)

K3074	47,XY,+16	34	29		Mat	Yes
K3084	47,XY,+16	34	30	Mat I or Pat I, II	Mat	Yes
K3086	47,XX,+16	32	30		Mat	Yes
K3164	47,XX,+16	35	33		Mat	Yes
K3259	47,XY,+16	33	29	Mat I, II or Pat I	Mat	Yes
K3361	47,XY,+16	34	31		Mat	
K3375	47,XY,+16	28	29	Mat I or Pat I, II	Mat	Yes
K3407	47,XX,+16	23	22		Mat	Yes
K3409	47,XX,+16	35	32	Mat I		
K3425	47,XY,+16	37	36		Mat	Yes
K3349	47,XY,+21	36	38	Mat I or Pat I, II	Pat	
K3023	47,XX,+21	37	36	Mat I		
K3175	47,XY,+21	48	35	Mat I		No#
K3275	47,XX,+21	37	36	Mat I or Pat I	Mat	Yes
K3403	48,XX,+7,+21	27	29	Mat II or Pat I	Mat	Yes

(Continued)

Table 3. (cont'd.)

K3034	46,X,+22	37	36	Mat I or Pat I	Mat
K3217	47,XY,+22	33	35	Mat I	Mat
K3272	47,XY,+22	35	31	Mat I	
K3296	47,XY,+22	29	29	Mat I or Pat I, II	
K3323	47,XX,+22	33	32	Mat I or Pat I, II	
K3359	47,XX,+22	28	23	Mat I	Mat
K3400	47,XX,+22	28	33	Mat I	Mat
K3418	47,XX,+22	31	27		
K3450	47,XX,+22	25	42		Mat

* Cytogenetic results are shown for all cases in which it was possible to rule out at least one of the four meiotic errors; DNA results are shown for all cases in which it was possible to specify the parent of origin; all other cases were uninformative.

+ Yes = recombination observed; no = no evidence for recombination at four or more informative loci; all other cases uninformative or had only 1-3 informative loci.

K3029 and K3056 were each homozygous at four different loci for which the parent of origin was heterozygous; K3175 was heterozygous at four loci for which the parent of origin was heterozygous.

Trisomy 13: Six trisomy 13 spontaneous abortions were studied, one of which (K3362) was a double aneuploid. We were able to determine the parental origin of the additional 13 in all six cases, with four being maternally derived and two paternally derived. In K3362, the extra 13 was paternal, while analysis of X-linked RFLPs indicated the single X was maternal; therefore, the error(s) leading to both the trisomy and the monosomy involved the paternal chromosome.

In four of the six cases, we observed recombination between the centromere and a locus defined by a DNA marker, or between two DNA loci. In three of these cases (K3177, K3392, and K2766), the trisomy 13 originated from an error at meiosis I and, therefore, the non-disjunctional event could not have resulted from failure to pair and/or recombine.

Trisomy 16: Cytogenetic studies of parental origin were informative in only two of 16 cases, although in five other cases we were at least able to exclude one or more of the four meiotic stages. Analysis of DNA markers was much more useful (Fig. 1); in 13 instances we were able to specify the parent of origin. By combining the cytogenetic with the DNA studies, we were able to determine the parental origin in 15 of the 16 cases, with 13 (86.7%) being maternal and two (13.3%) paternal in origin.

The results of our analysis of recombination in the 13 maternally derived cases are summarized in Figure 2. In two of the three cases of maternal meiosis I origin (K3084, K3375), recombination must have occurred between the two non-disjoined chromosomes, since, in both trisomies, we observed DNA markers at which parental heterozygosity had been reduced to homozygosity. In the third trisomy of maternal meiosis I origin (K3409), heterozygosity was maintained at the only two informative loci and, therefore, there was no evidence for crossing-over.

In the remaining ten cases of maternal origin, we were unable to specify the meiotic stage of origin and could only assess crossing-over between successive DNA markers. Nevertheless, we observed recombinants in six of the ten cases (K3074, K3086, K3164, K3259, K3407, K3425), indicating that the additional chromosome did not result from absence

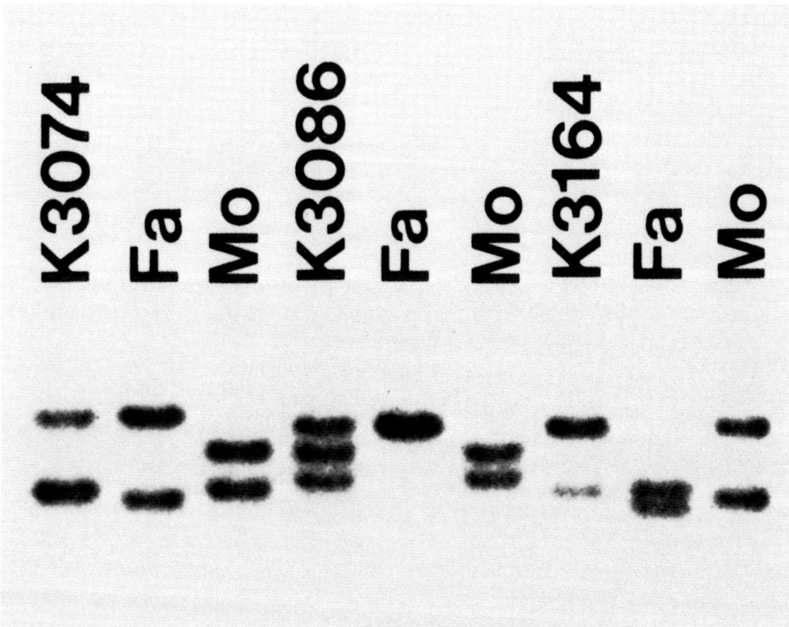

Figure 1. Analysis of non-disjunction in three trisomy 16 spontaneous abortions. DNA from the three cases and their parents were probed with pEKMDA2.1 (D16S83) which detects a multi-allelic polymorphism on 16p. In all three cases, the additional chromosome 16 was maternally derived; in two cases (K3074, K3164) the parental origin assignment was made on the basis of the relative intensity of the maternal and paternal alleles, while in K3068 the fetus clearly received two maternal alleles and one paternal allele.

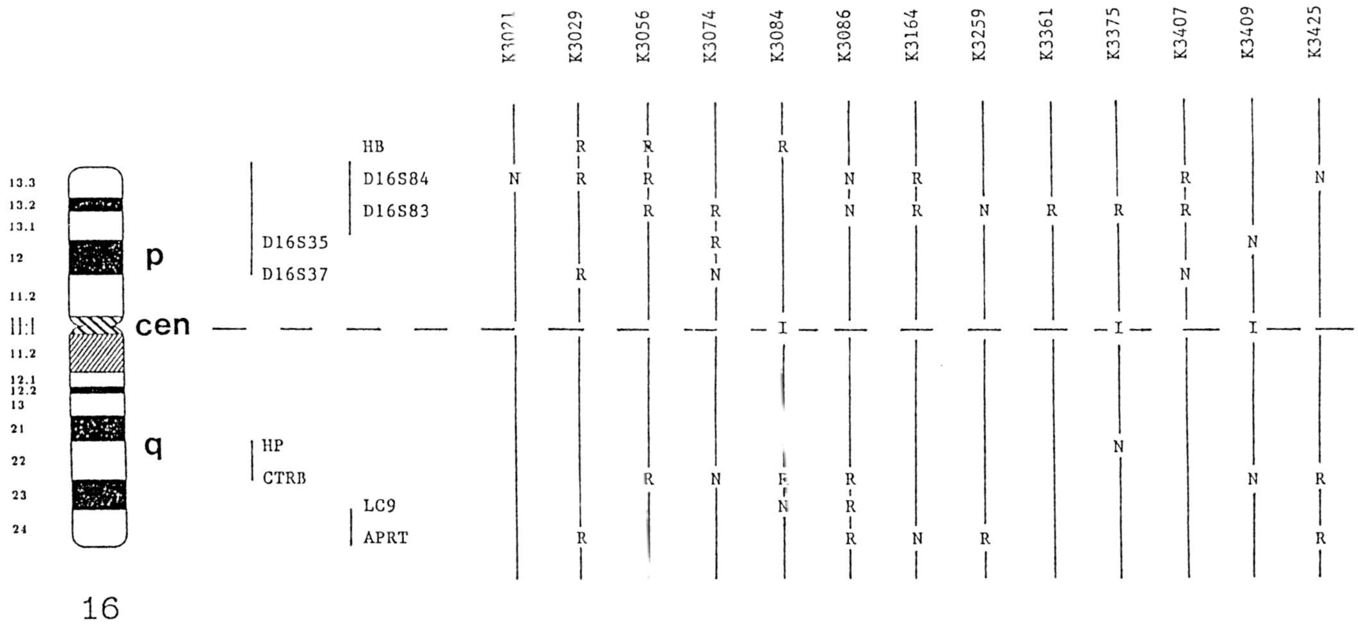

Figure 2. Analysis of recombination in 13 fetuses with an additional, maternally derived, chromosome 16. Centromeric chromosome heteromorphisms of the fetuses and mothers were evaluated and, in three cases (K3084, K3375, K3409) the results were consistent with an error at meiosis I (designated as "I"). All other cases were uninformative at the centromere. Additionally, trisomic fetuses were evaluated at any of nine DNA loci (HB, D16S84, D16S83, D16S35, D16S37, HP, CTRB, LC9, and APRT) heterozygous in the mother. Loci at which heterozygosity was maintained in the trisomic fetus were designated as "N", for not-reduced, and loci where heterozygosity was reduced to homozygosity were designated as "R".

Crossing-over between the centromere and a DNA marker is evident whenever a DNA loci designated as "R" is seen in association with meiosis I non-disjunction (e.g., K3084). Crossing-over between DNA loci is evident whenever one locus is designated as "R" and the other as "N" (e.g., K3074).

of pairing. Furthermore, in the only two cases (K3029, K3056) in which four or more informative loci were examined with no evidence for crossing-over, all loci were homozygous in the trisomic fetus. This observation is consistent with a mitotic origin for the trisomy, or with meiosis II non-disjunction and no crossing-over between the DNA loci, but is incompatible with meiosis I non-disjunction due to absence of pairing and/or recombination.

Trisomy 21: We studied five trisomy 21 spontaneous abortions, one of which (K3403) was trisomic for chromosome 7 as well as chromosome 21. We were able to determine the parental origin in all five cases, three on the basis of RFLPs and two on the basis of chromosome markers. Three of the five cases (K3023, K3175, K3275) were due to errors at maternal meiosis I, one (K3403) to an error at maternal meiosis II, and one (K3349) to an error at paternal meiosis I or II.

We identified recombinants in two of the five cases. However, in a third case (K3175) attributable to an error at maternal meiosis I, we were unable to demonstrate a cross-over between the centromere and four informative DNA markers.

Trisomy 22: Nine cases of trisomy 22 were studied and six were determined to be maternal in origin. In three cases we were unable to specify the parental origin. Two of the six assignments were made on the basis of RFLPs, one on the basis of cytogenetic analysis, and the other three were made on the basis of both DNA and chromosome markers.

One of the nine cases (K3034) was monosomic for the X chromosome as well as being trisomic for chromosome 22. However, analysis of X-linked RFLPs indicated that the monosomy was due to loss of the paternal X or Y while the extra 22 was maternal in origin. Therefore, the two errors were unrelated.

We were unable to detect recombination in any of the six cases of maternal origin. However, this is undoubtedly due to the limited number of loci thus far studied, as in three of the six cases none of the DNA markers were informative for recombination and, in the other three cases, only one or two DNA markers were informative.

DISCUSSION

The cytogenetic results of the present study demonstrate that maternal meiosis I non-disjunction is the most common source of autosomal trisomy. Thus, 83.3% (60/72) of cases of known parental and meiotic stage of origin were attributable to errors at this stage, compared to values of 9.7%, 5.5%, and 1.4% for errors at maternal meiosis II, paternal meiosis I, and paternal meiosis II, respectively. Furthermore, this effect was observed for all trisomies studied, with no trisomy having more paternally than maternally derived cases.

However, the cytogenetic observations also provide evidence for significant heterogeneity in origin among trisomies. Specifically, the observed level of paternal non-disjunction was 15 - 20% for trisomies 13 and 21, but less than 5% for the remaining trisomies. Very little additional information is available on the origin of trisomy 13, but trisomy 21 has been extensively studied with chromosome heteromorphisms, and information on parental origin is available for several hundred Down syndrome families (Hassold et al., 1984a). The observed level of paternal non-disjunction in these families is approximately 20%, i.e., a level at least as high as that observed in the present study. Thus, results from spontaneously aborted and liveborn trisomy 21 conceptuses indicate a higher than expected frequency of paternal errors, implying the existence of mechanisms of paternal non-disjunction largely, or entirely, restricted to chromosome 21.

However, the evidence for heterogeneity in origin among trisomies relies on analysis of chromosome heteromorphisms. This technique involves the subjective evaluation of size and staining intensity variants, and undoubtedly results in occasional misassignments of parental origin. Indeed, Carothers (1987) has suggested that the existence of a maternal age effect in trisomies of paternal origin, an effect observed in the present study and in most other cytogenetic studies of parental origin, is attributable to errors in parental origin assignments.

Therefore, we have been interested in using DNA markers to reassess the level of paternal non-disjunction in different autosomes. Our preliminary results, based on a series of 36 spontaneous abortions trisomic for chromosomes

13, 16, 21, or 22, are consistent with differing levels of paternal non-disjunction for different chromosomes. That is, considering only parental origin assignments made on the basis of DNA markers (with or without cytogenetic confirmation), we observed paternal non-disjunction in 40% (2/5) of cases of trisomies 13 and 21, but in only 6% (1/18) of cases of trisomies 16 and 22. The small number of cases studied makes it difficult to know the true significance of these observations, but the interpretation of different, chromosome specific frequencies of paternal non-disjunction is supported by results of other DNA marker studies of parental origin. Results from several such studies of trisomy 21 are summarized in Table 4, and these may be compared with similar data on trisomies 13, 16, and 22 from the present study, and with data on trisomy 18 from the study of Kondoh et al. (1988). Paternal errors were identified in only one of 20 cases of trisomies 16, 18, and 22, but accounted for approximately 20% of cases of trisomy 21. Thus, the relatively high incidence of paternal non-disjunction reported in cytogenetic studies of trisomy 21 appears to be real, and may be a characteristic feature of a limited number of chromosomes, including 21.

Table 4. Results of analyses of parental origin in autosomal trisomy based solely on DNA markers.

Trisomy	Study	PARENTAL ORIGIN Paternal %	 Maternal %
13	Present Study	1 (50%)	1 (50%)
16	Present Study	1 (8%)	12 (92%)
18	Kondoh et al. (1988)	0 (0%)	3 (100%)
21	Chakravarti et al.(this vol.) Rudd et al. (1988) Bricarelli et al. (1988) Galt et al. (in press) Buraczynska et al. (this vol.) Roulston et al. (this vol.)	28 (20%)	114 (80%)
22	Present Study	0 (0%)	5 (100%)

In addition to the analysis of parental and meiotic stage of origin of trisomy, we used RFLPs to study the relationship between errors of recombination and non-disjunction. To date, we have been primarily interested in evaluating the suggestion of Warren et al. (1987) that failure to pair is a causal factor in trisomy; therefore, we have scored trisomies as being positive for recombination (i.e., at least one cross-over observed) or negative for recombination (i.e., no cross-overs observed between four or more informative loci), with the latter category being consistent with a pairing failure etiology. The results of these analyses are summarized in Table 5 and from this it can be seen that failure to pair is relatively unimportant in the genesis of trisomies 13, 16, 21, and 22. That is, we observed recombination in 15 of 32 trisomies studied, and in most other cases there were little, or no, data available on recombination. Only three cases were negative for recombination using the above criteria and, of these, two were completely homozygous and could not have resulted from pairing failure. Thus, only one case, a trisomy 21, had data compatible with a pairing failure etiology.

These results are substantially different from those of Warren et al. (1987) and do not support their interpretation of a major role of pairing failure in the genesis of trisomy. However, the number of families evaluated in their study and in our own is still rather small, and the differences may be attributable to the limited data available. Alternatively, it may be that the effect of pairing failure is restricted to certain chromosomes and therefore not readily apparent in our series of different trisomies. For example, it might be that the effect is related to chiasma frequency, with those chromosomes having the fewest chiasmata being the most susceptible to pairing failure-related non-disjunction. This would be consistent with the available data on parental origin of trisomy, since the level of paternal non-disjunction is highest for 47,XXYs (Jacobs et al., 1988) and trisomy 21, situations in which the male bivalent is held together by one or at most two chiasmata. In any event, it is clear that a much more extensive series of trisomies, involving different chromosomes and different parental origins, must be examined before the possible role of recombinational errors in human non-disjunction is understood.

Table 5. Summary of observations on recombination in 32 trisomies in which the parental origin of trisomy was determined by DNA and/or cytogenetic markers.

Trisomy	Parent of Origin	Evidence for Recombination* Yes	No	Unknown
13	Paternal	1		1
	Maternal	3		1
16	Paternal	1		1
	Maternal	8	2[+]	3
21	Paternal			1
	Maternal	2	1[#]	1
22	Maternal			6
TOTAL	Paternal	2		3
	Maternal	13	3	11

* Yes = recombination observed; no = no evidence for recombination at four or more informative loci; all other cases were uninformative or had only 1-3 informative loci.

[+] Both cases were homozygous at four loci at which the parent of origin was heterozygous.

[#] Case was heterozygous at four loci at which the parent of origin was heterozygous.

ACKNOWLEDGEMENTS

We wish to express our gratitude to Jane Hersey for preparation of this manuscript and to the staff of the Pathology Department, Kapiolani-Children's Medical Center, Honolulu, for collection of the tissue samples, and to M. Leppert, Y. Nakamura, J. Holden, and S. Latt for donation of DNA probes. This research was supported by N.I.H. grants HD21341 and HD25509.

REFERENCES

Ayme S, Lippman-Hand A (1982). Maternal-age effect in aneuploidy: does altered embryonic selection play a role? Am J Hum Genet 39:558-565.

Bricarelli FD, Pierluigi M, Perroni L, Grasso M, Arslanian A, Sacchi N (1988). High efficiency in the attribution of parental origin of non-disjunction in trisomy 21 by both cytogenetic and molecular polymorphisms. Hum Genet 79:124-127.

Buraczynska M, Stewart GD, Sherman S, Freeman V, Grantham M, Uchida I, Hassold T, Kurnit DM. Studying nondisjunction of chromosome 21 with cytogenetic markers on the short arm and DNA markers encompassing the long arm. (this volume)

Chakravarti A, Majumder PP, Slaugenhaupt SA, Surti U, Ferrell RE, Warren AC, Antonarakis SE. Gene centromere mapping and the study of non-disjunction in autosomal trisomies and ovarian teratomas. (this volume)

Degau JH, van de Sande JH, Lin CC (1978). $(CMA)_2S$: a double intercalating analogue of quinacrine for chromosome banding. Am J Hum Genet 30:78A.

Feinberg AP, Vogelstein B (1983). A technique for radiolabeling restriction fragments to high specific activity. Anal Biochem 132:6-13.

Galt J, Boyd E, Conner JM, Ferguson-Smith MA. Isolation of chromosome 21 specific DNA probes and their use in the analysis of nondisjunction in Down syndrome. Hum Genet (in press).

Hassold T, Chen N, Funkhouser J, Jooss T, Manuel B, Matsuura J, Matsuyama A (1980). A cytogenetic study of 1,000 spontaneous abortions. Ann Hum Genet 44:151-178.

Hassold, T, Chiu D, Yamane J (1984a). Parental origin of autosomal trisomies. Ann Hum Genet 48:129-144.

Hassold T, Jacobs PA, Leppert M, Sheldon M (1987). Cytogenetic and molecular studies of trisomy 13. J Med Genet 24:725-732.

Hassold T, Kumlin E, Leppert M, Takaesu N (1985). Determination of the parental origin of sex chromosome monosomy using restriction fragment length polymorphisms. Am J Hum Genet 37:956-972.

Hassold T, Warburton D, Kline J, Stein Z (1984b). The relationship of maternal age and trisomy among trisomic spontaneous abortions. Am J Hum Genet 36:1349-1356.

Jacobs PA, Hassold TJ, Henry A, Pettay D, Takaesu N (1987). Trisomy 13 ascertained in a survey of spontaneous abortions. J Med Genet 24:721-724.

Jacobs PA, Hassold TJ, Whittington E, Butler G, Collyer S, Keston M, Lee M (1988). Klinefelter's syndrome: an analysis of the origin of the additional sex chromosome using molecular probes. Ann Hum Genet 52:93-109.

Kondoh T, Tonoki H, Matsumoto T, Tsukahara M, Niikawa N. Origin of the extra chromosome in trisomy 18 (1988). Hum Genet 79:377-378.

Merriam JR, Frost JN (1964). Exchange and nondisjunction of the X-chromosomes in female Drosophila melanogaster. Genetics 49:109-122.

Roulston D, Antonarakis SE, Lewis JG, Cohen MM, Schwartz S. Cytological and molecular studies of nucleolar organizing region variants and recombination in trisomy 21. (this volume)

Rudd NL, Dimnik LS, Greentree C, Mendes-Crabb K, Hoar DI (1988). The use of DNA probes to establish parental origin in Down syndrome. Hum Genet 78:175-178.

Stewart GD, Hassold TJ, Berg A, Watkins P, Tanzi R, Kurnit DM (1988a). Down syndrome: studying nondisjunction and meiotic recombination using cytogenetic and molecular polymorphisms that span chromosome 21. Am J Hum Genet 42:227-236.

Surosky RT, Tye B-K (1988). Meiotic disjunction of homologs in Saccharomyces cerevisiae is directed by pairing and recombination of the chromosome arms but not by pairing of the centromeres. Genetics 119:273-287.

Warren AC, Chakravarti A, Wong C, Slaugenhaupt SA, Halloran SL, Watkins PC, Metaxotou C, Antonarakis SE (1987). Evidence for reduced recombination on the nondisjoined chromosome 21 in Down syndrome. Science 237:652-654.

Molecular and Cytogenetic Studies of
Non-Disjunction, pages 135–151

THE ORIGIN OF SEX CHROMOSOME ANEUPLOIDY

Patricia Jacobs*, Terry Hassold[0], John Harvey* and Kristen May[0]

*Wessex Regional Cytogenetic Unit S.G.I. Salisbury, Wilts U.K. SP2 7SX, and [0]Dept. of Pediatrics, Emory University, School of Medicine Atlanta, Georgia 30322

INTRODUCTION

Approximately 1 in every 500 newborn babies has an abnormal sex chromosome constitution. In the great majority this is due to the presence of an additional X or Y chromosome while in a small minority it is due to the absence of a second sex chromosome or the presence of a structurally abnormal sex chromosome (Hook and Hamerton, 1977). Among fetal deaths however, the situation is quite different with no fewer than 10% having a sex chromosome abnormality. In the vast majority this is due to the absence of a second sex chromosome and in only in a minority is it due to presence of an additional X chromosome (Hassold and Jacobs, 1984) (Table 1) If we assume that 15% of all pregnancies that survive long enough to be clinically recognised abort spontaneously and that a further 1% are stillborn, we can estimate that 1.3% of all clinically recognised conceptions have a missing sex chromosome and that 0.2% have an additional sex chromosome. A comparable figure for conceptions with an additional chromosome 21 is 0.45%. Thus while the presence of an additional chromosome 21 is much more frequent at conception than the presence of an additional sex chromosome, neither occur with anything like the frequency of monosomy X. The relative probability of survival to birth is vastly different for the different aneuploidies, ranging from a low of 0.3% for monosomy X through 22% for trisomy 21, 55% for the XXY condition,

70% for the XXX condition to a high of 100% for XYY males. It is, therefore, well to remember when investigating the etiology of chromosome aneuploidy that the liveborn may represent only a very small, and possibly unrepresentative, sample of the abnormality under study.

TABLE 1.

FREQUENCY OF CHROMOSOME ABNORMALITIES

	Abnormalities %				
Population	45,X	47,XXY	47,XXX	47,XYY	47,+21
Spontaneous Abortions	-8.6	0.2	0.1	--	2.3
Stillbirths	0.25	0.4	0.3	--	1.3
Liveborn	<0.01	0.05	0.05	0.05	0.13
All Clinically Recognized Pregnancies*	1.30	0.08	0.06	0.04	0.45
Relative Probability of Survival to Birth %	0.3	55.3	70	100	22.2

* assume 15% spontaneous abortion, 1% stillbirth

While sex chromosome aneuploidy is a very common event in our species little is known about the causal factors. The first extensive use of a gene to detect the parental origin of a human aneuploidy was the study by Sanger *et al* (1977) of the segregation of the X linked blood group gene Xg in liveborn individuals with sex chromosome aneuploidy and their parents. These authors reported that approximately 25% of 45,X individuals had a paternal X and the remainder a maternal X and that some 40% of 47,XXY individuals resulted from an error in spermatogenesis while 60% were maternal in origin. However, it is not possible to differentiate between one, two or three doses of the Xg^a antigen and thus very few matings are fully informative and results must be based on maximum likelihood estimates. Furthermore the Xg^a system provides no information on the origin of 47,XXX females.

The recent development of large numbers of DNA probes that detect restriction fragment length polymorphisms on the human chromosomes provides an opportunity to investigate the precise origin of virtually all sex chromosome abnormalities. We are currently engaged in determining the origin of human sex chromosome abnormalities using such molecular probes. We are interested in determining the parent and, where possible, the cell division at which the error resulting in the sex chromosome aneuploidy occurred; constructing centromere maps of the X chromosome to determine whether non-disjunction at the first meiotic division is associated with an abnormal level or distribution of recombinant events and in determining whether the parental origin of the sex chromosome aneuploidy affects the phenotype of the resulting conceptus.

Our preliminary results on the origin of the 45,X, 47,XXY and 47,XXX constitutions are summarized in this paper and compared to similar information on trisomy 21.

SEX CHROMOSOME MONOSOMY

X-chromosome monosomy can arise as a result of a first or second meiotic division error during gametogenesis in either the father or the mother, or as a result of mitotic nondisjunction or anaphase lag in an early division of a chromosomally normal zygote with subsequent loss or exclusion from the embryo proper of the normal cell line. In the 45,X conceptions all that is present is a single X that did <u>not</u> go wrong. The parental origin of this non-aberrant sex chromosome can be identified and it can be assumed that the error occurred during gametogenesis in the other parent or at an early cell division of the zygote and involved the chromosome of the "missing" parent. However it is not possible to determine the cell division or mechanism which gave rise to the error.

TABLE 2.

PARENTAL ORIGIN OF 45,X

POPULATION	X^M No.	X^M %	X^P No.	X^P %	TOTAL No.
Spontaneous abortions	27	79.4	7	20.6	34
Livebirths	12	92.3	1	7.7	13
Total	39	83.0	8	17.0	47

We have data on the parental origin of 47 sex chromosome monosomies, 27 of which are spontaneous abortions and 12 live births (Table 2). Of these 39 have an X^M and 8 an X^P constitution, a proportion in good agreement with that obtained with Xg^a blood groups in liveborn 45,X females. However when the data are considered separately for spontaneous abortions and live births there is a statistically non significant suggestion of a difference, there being a smaller proportion of X^P cases in the live births. Furthermore the single X^P among the live births was a very severely affected baby born prematurely who required many months in an intensive care unit. If our initial suggestion of a deficiency of the X^P class among live births is substantiated after more cases are tested, it will be necessary to explain the discrepancy between the results obtained by Xg blood groups on live births and those obtained using molecular probes. One possible explanation might be that many of the live births tested with Xg^a were cryptic mosaics, whereas both X and Y linked probes were used in the molecular study in an effort to exclude such mosaics. It is known that in recognised 45,X/46,X marker mosaics the single X may be paternal in origin.

A consideration of parental age in the 45,X monosomies confirms the previous observation of a significantly reduced maternal age in the 45,X^P class (Table 3) (Hassold et al, 1988).

TABLE 3.

PARENTAL AGE OF 45,X

PARENT OF ORIGIN	POPULATION	PATERNAL AGE (N)	MATERNAL AGE (N)
X^M	Spontaneous abortions	32.9 (27)	29.7 (27)
	Live Births	30.6 (8)	27.1 (12)
	All X^M	32.4 (35)	28.9 (39)
X^P	Spontaneous abortions	27.9 (7)	24.4 (7)
	Live Births	20 (1)	15 (1)
	All X^P	26.9 (8)	23.2 (8)

47,XXY

The ways in which a 47,XXY individual could arise are shown in Table 4. Because the product of the error is present it is possible, if enough appropriate probes are used, to discriminate among all the mechanisms shown in

TABLE 4.

MECHANISM OF ORIGIN OF 47,XXY MALES

ORIGIN OF EXTRA X	CELL DIVISION	CENTROMERE	MECHANISM	X-LINKED LOCI
PATERNAL	Mei I	--	Nullo-chiasmate ND	Identical to father
			Chiasmate ND	Same as father except for distal pseudo-autosomal loci that will differ in 50%
MATERNAL	Mei I	Heterozygous	Nullo-chiasmate ND	Identical to mother
			Normo-chiasmate ND) Hypo-chiasmate ND) Hyper-chiasmate ND)	Proximal heterozygous Distal homozygous
	Mei II	Homozygous	Non-disjunction	Proximal homozygous Distal heterozygous
	Postzygotic mitotic	Homozygous	Non-disjunction	Completely homozygous

the table. We have published data on 39 47,XXY males (Jacobs et al, 1988) and the results on these together with a further 26 cases are shown in Table 5. The parent in whom the error occurred was determined in 59 cases and in the remaining 6 cases DNA was available from the proband and mother but not from the father. In all 6 a large number of probes were tested and no alleles seen in the proband that were not present in the mother. It is therefore highly likely that all 6 were maternal in origin, and they have been considered as such in our analyses. As can be seen 33 of the cases were paternal in origin and 32 maternal, confirming our previous observation that approximately 50% of 47,XXY males arise as a result of an error of the first paternal meiotic division, a very high proportion by comparison with all other studied trisomies. We tested 13 paternally derived 47,XXY individuals, with pseudoautosomal probes (Page et al, 1987). In 9 cases the Klinefelter was identical to the father while 4 showed evidence of recombination (Page et al unpublished). A single crossover in the pseudoautosomal region will result in one recombinant and one nonrecombinant XXY, thus our observations show that at least 8 of the tested cases arose subsequent to a crossover event. These preliminary data suggest that the majority of paternally derived Klinefelters arise as a result of chiasmate nondisjunction, i.e. normal pairing and exchange takes place between the sex chromosomes prior to nondisjunction, and only a minority arise from a spermatocyte in which there is no recombination between the X and Y chromosomes.

TABLE 5.

PARENTAL ORIGIN OF 47,XXY MALES

PARENT	NO.		%
Paternal	33		51
Maternal	26	40)	49
Not determined	6	9)	
TOTAL	65		

Among the 32 maternally derived 47,XXY individuals we were able to determine the centromere status in 23 (Table 6); 18 were compatible with a meiosis I error and 5 with a meiosis II error. In 5 of the 18 cases resulting from a meiosis I error there was no evidence of reduction of any maternal heterozygous loci (Table 7). However in 2 of these there were only 1 and 2 informative maternal loci respectively. In the remaining 3 there were 9 or more informative maternal loci tested with no evidence of reduction in the 47,XXY offspring, suggesting that at least 3 out of the 18 resulted from nullo-chiasmate non-disjunction.

TABLE 6.

CELL DIVISION OF ORIGIN OF 47,XXY (MATERNAL)

PATIENTS	MEI I	MEI II	?	TOTAL
Confirmed Maternal	15	4	7	26
Probably Maternal	3	1	2	6
TOTAL	18	5	9	32

TABLE 7.

47,XXY MALES OF MATERNAL MEI I ORIGIN

PATIENT	INFORMATIVE PROBES* REDUCED/NONREDUCED	COMPATIBLE WITH NULLO-CHIASMATE NONDISJUNCTION
88/662	3/0	N
PJ153	4/0	N
PJ163	3/2	N
Y2	5/3	N
Y24	0/12	Y
Y54	0/9	Y
Y55	0/12	Y
PJ24	3/6	N
PJ25	3/1	N
PJ40	0/2	Y
PJ68	1/5	N
PJ77	2/2	N
PJ97	1/2	N
SI636	2/3	N
X183	0/1	Y
PJ3°	1/4	N
PJ5°	2/2	N
Y76°	1/1	N

* Mother heterozygous - excluding contromere probe

° Presumed maternal

TABLE 8.

47,XXY MALES OF MATERNAL MEI II ORIGIN

PATIENT	INFORMATIVE PROBES* REDUCED/NONREDUCED	COMPATIBLE WITH POSTZYGOTIC MITOTIC ERROR
88/2830	2/0	Y
PJ92	6/0	Y
PJ91	1/5	N
PJ54	2/2	N
PJ96[0]	2/1	N

* Mother heterozygous - excluding centromere probe

0 Presumed maternal

Among the five 47,XXY cases compatible with a meiosis II error, there were 2 in which all informative probes were reduced, an observation compatible with a postzygotic mitotic error (Table 8). However in 1 case there were only two informative probes while in the other there were 6. We consider the latter case as probably resulting from a postzygotic mitotic error, to date the only such contender in our series.

TABLE 9.

PARENTAL AGE OF 47,XXY MALES

ORIGIN	PATERNAL AGE (N)	MATERNAL AGE (N)
Paternal	30.0 (26)	27.1 (28)
Maternal	32.6 (27)	31.0 (28)
Mei I	33.5 (15)	32.5 (16)
Mei II	28 (4)	27.0 (4)
?	33.3 (8)	30.5 (8)

The parental ages of the different classes of 47,XXY individual are shown in Table 9. Among the paternally derived XXY's the parental ages are unremarkable and very similar to those of a previously reported control series (Jacobs et al, 1988). The maternally derived cases are associated with an increased maternal age and, on the small number tested to date, this increase appears to be confined to those individuals of maternal meiotic I origin.

TABLE 10.

MECHANISM OF ORIGIN OF 47,XXX FEMALES

ORIGIN OF EXTRA X	CELL DIVISION	CENTROMERE	MECHANISM	X-LINKED LOCI
PATERNAL	Mei II	Homozygous	Nondisjunction	Distal pseudoautosomal loci heterozygous in 100%
	Postzygotic mitotic	Homozygous	Nondisjunction	Completely homozygous Same as father except for distal pseudo-autosomal loci that will differ in 50%
MATERNAL	Mei I	Heterozygous	Nullo-chiasmate ND	Identical to mother
			Normo-chiasmate ND) Hypo-chiasmate ND) Hyper-chiasmate ND)	Proximal heterozygous Distal homozygous
	Mei II	Homozygous	Nondisjunction	Proximal homozygous Distal heterozygous
	Postzygotic mitotic	Homozygous	Nondisjunction	Completely homozygous

47,XXX FEMALES

The possible mechanisms of origin of 47,XXX females are shown in Table 10. Again, as in the case of 47,XXY males, because the product of the abnormal event is present it is possible, if sufficient probes are used, to discriminate among all possible mechanisms of origin. We have studied 15 such females and found 1 to be paternal in origin, the remaining 14 being maternal (Table 11). The cell division of origin has been established in 8 of the 14 maternal cases (Table 12), with 6 being attributable to meiotic I errors and 2 to meiotic II errors. In one of the meiotic I cases there were 7 tested loci at which the mother was heterozygous and none were reduced in the 47,XXX female, so this case may well be an example of nullo-chiasmate nondisjunction. In neither of the meiotic II cases were the results compatible with a mitotic error.

TABLE 11.

PARENTAL ORIGIN OF 47,XXX FEMALES

PARENT	NO.	%
Paternal	1	6.7
Maternal	14	93.3
TOTAL	15	

TABLE 12.

CELL DIVISION OF ORIGIN OF 47,XXX (MATERNAL)

PATIENTS	MEI I	MEI II	?	TOTAL
Maternal	6	2	6	14

TABLE 13.

PARENTAL AGE OF 47,XXX FEMALES

ORIGIN	PATERNAL AGE (N)	MATERNAL AGE (N)
Paternal	36 (1)	30 (1)
Maternal	29.8 (6)	28.7 (11)
Mei I	31.3 (3)	30.0 (6)
Mei II	25 (1)	23.2 (2)
?	30 (2)	29.7 (3)

The parental ages of the 47,XXX females are shown in Table 13. The numbers are too small to draw any conclusions but they may be similar to maternally derived 47,XXY males, with an increased maternal age effect being restricted to the cases of maternal meiotic I origin.

EFFECT OF RECOMBINATION ON NONDISJUNCTION OF THE SEX CHROMOSOMES

We have as yet too little information on levels of recombination in sex chromosomes undergoing non-disjunction at the first meiotic division to know what role, if any, abnormalities of recombination play in non disjunction. Among 24 cases of 47,XXY and 47,XXX patients of maternal meiotic I origin our observations were compatible with nullo-chiasmate nondisjunction in 4, suggesting that complete failure of recombination may be associated with nondisjunction in some 16% of cases.

We also have preliminary evidence suggesting that an increased level of recombination may be associated with nondisjunction of the X chromosome at the first maternal meiotic division. Recently we have been testing all sex chromosome aneuploids with probe M27B, kindly provided by Dr Ian Craig of the Genetics Department, Oxford University. This probe detects multiple X-linked restriction fragments and is on the short arm of the X some 8 cM from the centromere (Dr Ian Craig, personal communication). In spite of the relative closeness of this probe to the centromere, we have 6 recombinants out of 14 maternal meiotic I derived 47,XXY and 47,XXX individuals, including the 4 who may be the result of nullo-chiasmate non disjunction. These data suggest that there is an increase of recombination, at least in the proximal region of X^p, associated with maternal meiotic I chiasmate nondisjunction. This situation is reminiscent of that in Drosophila melanogater where X chromosome non-disjunction has been shown to be associated with increased levels of recombination near the centromere (Merriam and Frost, 1964).

COMPARISON OF ORIGIN OF SEX CHROMOSOME ABNORMALITIES AND TRISOMY 21

The data on the origin of sex chromosome aneuploids and on trisomy 21 are given in Table 14. It has been assumed that all such cases result from meiotic errors and the data are given by parent and cell division of origin expressed as a percentage of all clinically recognised conceptions. If the 45,X category is disregarded, it can be seen that paternal meiotic I errors are equally frequent for the sex chromosomes and chromosome 21. Paternal meiotic II errors occur with equal frequency for the Y and 21 while paternal meiotic II errors involving the X appear to occur with only about one tenth the frequency. Thus, in terms of paternal nondisjunction, chromosomes 21 and the Y are remarkably similar.

TABLE 14.

FREQUENCY AND ORIGIN OF CHROMOSOME ABNORMALITIES AMONG ALL CLINICALLY RECOGNISED CONCEPTIONS (%)

ABNORMALITY	XO	XXY	XXX	XYY	+21
All	1.5	0.08	0.06	0.04	0.45
Paternal error	I II	I II	I II	I II	I II
	1.2	0.04 -	- 0.004	- 0.04	0.058 0.031
Maternal error	I II	I II	I II	I II	I II
	0.3	0.031 0.009	0.04 0.014	- -	0.31 0.058

In females nondisjunction of the sex chromosomes at the first meiotic division occurs about twice as frequently as in the male and is equally likely to give rise to 47,XXY and 47,XXX conceptuses. However, maternal meiotic I nondisjunction of chromosome 21 appears to be about four times as common as for the X chromosome. Allowing for the very small number of cases, it appears that second meiotic nondisjunction is, at most, about a quarter as frequent as first maternal meiotic non-disjunction and equally likely to give rise to 47,XXY and 47,XXX conceptuses. Errors of the second maternal meiotic division involving chromosome 21 appear to be about three times as common as those involving the X chromosome. Thus in females chromosome 21 seems to have a much greater propensity than the X chromosome for non-disjunction at both the first and second meiotic divisions.

ACKNOWLEDGEMENTS

We are extremely grateful to all the individuals who provide us with specimens for study and to Muriel Lee, Dorothy Pettay and Norman Takaesu for their excellent technical assistance. This work was supported by grant HD 21341 from the National Institute of Health.

REFERENCES

Hassold T, Benham F, Leppert M (1988). Cytogenetic and molecular analysis of sex chromosome monosomy. Am J Hum Genet 42:534-541.

Hassold T, Jacobs P A (1984). Trisomy in man. A Rev Genet 18:69-97.

Hook E B, Hamerton J L (1977). The frequency of chromosome abnormalities detected in consecutive newborn studies. In Population Genetics (ed. E B Hook and I H Porter) New York: Academic Press pp 63-79.

Merriam J R, Frost J N (1964). Exchange and Nondisjunction of the X Chromosomes in Female Drosophila melanogaster. Genetics 49:109-122.

Page D C, Bieker K, Brown L G, Hinton S, Leppert M, Lalouel J, Lathrop M, Nystrom-Lahti M, De La Chapell A, White R (1987) Linkage, Physical Mapping, and DNA Sequence Analysis of Pseudoautosomal Loci on the Human X and Y Chromosomes. Genomics 1:243-256.

Sanger R, Tippett P, Gavin J, Teesdale P, Daniels G L (1977). Xg groups and sex chromosome abnormalities in people of northern European ancestry: an addendum. J Med Genet 14:210-213.

Molecular and Cytogenetic Studies of
Non-Disjunction, pages 153–163

TISSUE SPECIFIC TRISOMY - THE SIGNIFICANCE OF CONFINED PLACENTAL MOSAICISM.

Dagmar K. Kalousek

Department of Pathology, University of British Columbia, Vancouver, B.C., Canada.

ABSTRACT

Confined placental mosaicism is tissue specific chromosomal mosaicism arising in both chromosomally normal and abnormal zygotes. Studied by first trimester chorionic villus sampling it has been reported in about 1-2% of human gestations. It is more common in the cytotrophoblast than in placental connective tissue. While intrauterine survival of some chromosomally abnormal embryos and fetuses appears to be significantly enhanced by its presence, the effect of confined placental mosaicism on chromosomally normal fetuses is variable and at present time still unpredictable.

INTRODUCTION

Chromosomal trisomies represent the most frequently detected numerical chromosomal abnormality in humans (Jacobs and Hassold, 1987). Trisomy for any chromosome can occur in human gametes (Plachot et al, 1988 and Martin et al, 1986), but the viability of the developing trisomic conceptus is highly compromised. Although trisomy for all chromosomes has been reported in preimplantation losses and spontaneous abortuses (Jacobs and Hassold, 1987, Watt et al, 1988) only very few trisomic conceptions survive to term. Thus, chromosomal trisomies in human conceptions have one feature in common; they usually act as a lethal mutation for embryonic and fetal development. Rarely malformed infants with non-mosaic trisomy 8, 9 and 22 have been described surviving to term (de Grouchy and Turleau, 1982). Of conceptions with trisomy 13 and 18 less than 5% are liveborn. For trisomy 21 the figure is 20-35% (Bond and Chandley, 1983; Jacobs and Hassold, 1987).

In trisomic conceptions, mosaicism involving both embryo/fetus and placenta is infrequent. Among trisomic liveborn infants, only 1-2% with trisomy 21 and fewer with trisomy 13 and 18 are mosaic. However, confined placental mosaicism has been shown to be present in all viable conceptions with non-mosaic trisomy 13 and 18 after 20 weeks of gestation. In these cases confined placental mosaicism is due to a postzygotic loss of a trisomic chromosome in a progenitor cell of the trophectoderm resulting in tissue specific diploid/trisomy mosaicism, while the rest of the conceptus shows non-mosaic trisomy (Kalousek et al, 1989).

Detection of confined mosaicism in placental tissue is not limited to conceptions developing from a trisomic zygote, it has also been reported in 1-2% of viable diploid conceptions studied with first trimester chorionic villus sampling (CVS) (Mikkelson and Ayme, 1987). In these pregnancies confined placental mosaicism can assume 2 main forms:

a) aneuploidy in pure or mosaic form in the cytotrophoblast, associated with chromosomally normal placental stroma and fetus (Type I).
b) diploidy in the cytotrophoblast and embryo/fetus, associated with mosaic aneuploidy in placental stroma (Type II).

Type III confined placental mosaicism is similar to type I, but occurs in aneuploid conceptions. Thus diploidy in pure or mosaic form in the cytotrophoblast is associated with non-mosaic aneuploidy in placental stroma and embryo/fetus.

Although confined or tissue specific mosaicism may also exist in the embryo or fetus as documented by detection of aneuploid cell clones in connective tissue of some dysmorphic or mentally retarded individuals

(Pagon et al, 1974; Peltomaki et al, 1987) the discussion of this topic is beyond the scope of this report.

In this paper, the present knowledge about placental tissue cytogenetics is summarized and correlated with intrauterine survival of the embryo and fetus. The conclusions are based on both published results and those collected in the cytogenetic research laboratory at the B.C. Children's Hospital.

MATERIALS AND METHODS

Collection of tissue

Placental tissue was collected from spontaneous abortuses, from viable first trimester pregnancies studied by chorionic villus sampling (CVS), from products of second trimester pregnancies terminated for prenatally diagnosed aneuploidy and from term placentas.

Spontaneous abortions were defined as any spontaneously aborted conceptions or those conceptions surgically removed after ultrasound diagnosis of missed abortion, blighted ovum or empty sac. The gestational age varied from 8-14 weeks.

In pregnancies studied by CVS the chorionic villi were obtained by transcervical aspiration using a Portex catheter under ultrasound guidance at 10-11 weeks of gestation. The main indication for CVS was advanced maternal age. Only a small number of mothers had the procedure done for diagnosis of biochemical or DNA defects.

Second trimester and term placentas were obtained after delivery of the fetus. Only placentas from pregnancies known to be aneuploid or presenting as unexplained intrauterine growth retardation (IUGR) were collected. The weight criteria for diagnosis of IUGR using both sex and gestational age were based on statistics derived from a Canadian population by Stats Canada in the period 1980-1984. In pregnancies with IUGR, gestational age was confirmed by ultrasound in 95% of cases, while only menstrual dates were available in the remainder.

<u>Cytogenetic techniques</u>

Cultures of peripheral blood lymphocytes were established from cord blood and harvested according to standard cytogenetic techniques. Several explants of the chorionic villi, chorion and amnion were seeded on plastic petri dishes containing nutrient media and incubated at 37°C. After 2-3 weeks cells were harvested in situ. Direct preparations from chorionic villi obtained were prepared using a modified methodology of Simoni et al, 1983 and 24-48 hour incubation (Vekemans and Perry, 1984). When intact placentas or chorionic sacs were available, multiple sites (including cord insertion area and two peripheral sites) were sampled for each of the tissues . In cases of pregnancy termination, in addition to placental tissues, multiple fetal tissues were cultured and cytogenetically analyzed.

All chromosomal analyses were done after G-banding. At least 15 cells were analyzed for each tissue. A preparation was considered to exhibit true mosaicism if 3 or more cells had an identical chromosomal complement, but differed from the basic constitutional karyotype. In cases of identified diploid/aneuploid mosaicism in direct preparations, sex chromosomes could be used to rule out maternal contamination since half of the newborns and fetuses were males. As no maternal contamination has ever been seen in direct preparation in this laboratory, no chromosomal polymorphism studies were initiated in cases of female sex and diploid/aneuploid mosaicism. The diagnosis of chromosomal mosaicism in long term cultures required the minimum of 3 cells be identified in more than one dish to rule out the evolution of an <u>in vitro</u> cell line.

RESULTS

To evaluate confined placental mosaicism, the results from both direct preparations from cytotrophoblast and long term cultures of chorionic villi, chorion or amnion are used.

Thirty first trimester spontaneous abortions have been studied and 3 mosaic cases have been identified showing dichotomy between cytotrophoblast and cultured tissues. Two cases involved diploid/tetraploid mosaicism. In both cases the diploid cell line was observed in the direct preparation the and tetraploid line in all long term cultures initiated from different placental tissues: amnion, umbilical cord, chorion, and villous stroma. The third case revealed mosaicism for trisomy 4 in all placental cultures while the cytotrophoblast and embryo proper showed only normal diploid karyotype.

Two different types of confined placental mosaicism were identified among 7 mosaics detected in 243 cases of CVS. In 5 cases the mosaicism involving an aneuploid cell line either in mosaic (2 trisomy 13, one triple trisomy 6, 7, 20) or pure form (trisomy 12 and extra marker chromosome) was confined to the cytotrophoblast. In 2 cases mosaicism was limited to villous stroma (trisomy 18 and monosomy X). In all cases but one, in which the family decided to terminate the pregnancy, follow up amniocentesis revealed a normal diploid karyotype and all newborns were clinically normal (Kalousek et al, 1987). In addition, long term cultures initiated after term delivery from different placental tissues also revealed only normal diploid complements.

The results of cytogenetic analysis of 28 placentas from viable trisomic fetuses and newborns revealed that in all cases of non-mosaic fetal/newborn trisomy 13 and trisomy 18, mosaicism involving a diploid/aneuploid cell line was identified in the cytotrophoblast. No mosaicism in any other placental/ fetal tissues or cord blood was detected. Although the percentage of cells with a normal karyotype in the cytotrophoblast varied greatly (12-100%), normal diploid cells were present in all preparations. In the 12 cases of trisomy 21 no evidence of placental mosaicism was found (Kalousek et al, 1989).

The cytogenetic results of 74 placentas from pregnancies with unexplained fetal IUGR revealed 2 cases of confined chorionic mosaicism among 62 placentas processed for long term cultures only (Kalousek and Dill, 1983). Twelve placentas processed for both direct preparations and long term cultures yielded one mosaic placenta involving a 45, X cell line confined to the cytotrophoblast.

DISCUSSION

Chromosomal mosaicism originates in early embryonic development through nondisjunction, anaphase lag or structural rearrangement. The resultant mosaic pattern in the conceptus depends on many factors: the number of blastomeres at the time of mutational event, the cell lineage affected by the mutational event and cell selection based on viability of mutant cells.

The existence of confined placental mosaicism restricted either to the cytotrophoblast, the extra-embryonic mesoderm, or to both of these lineages, with a complete absence of mosaicism in the embryo, reflects the complexity of both early placental development and differential viability of mutant cells in different lineages. Confined placental mosaicism thus results from viable chromosomal mutation occurring in progenitor cells of trophoblast or extraembryonic mesoderm during morula or blastocyst stage and not affecting a small number of embryoblasts from which development of the embryo proper is initiated. (Gardner, 1978; Markert and Peters, 1978; Kalousek, 1985).

Although in chorionic villi three different lineages, trophoblast, extraembryonic mesoderm and embryonic mesoderm are represented, the lineage containing abnormal cell line can be easily traced based on its location in villi and its correlation with embryonic/fetal karyotypes. During placental development cells of the trophoblast form primary villi with the syncytiotrophoblast externally located and the cytotrophoblast internally situated. Secondary villi develop by migration of cells from both extra-embryonic mesoderm and primitive embryonic streak into the villous core. Tertiary villi are characterized by the appearance of primitive capillaries in the villous core around day 18. Tertiary villi represent the final villous structure and are derived from three different cell lineages: polar trophectoderm, extra-embryonic mesoderm and primitive embryonic streak. The trophoectoderm lineage is analyzed by direct preparation or by a short term incubation of chorionic villi. Fibroblasts in long term cultures of chorionic villi and chorion plate represent the extraembryonic and embryonic mesoderm lineages.

Although the precise role of confined placental mosaicism in intrauterine development cannot be determined yet, some data suggest it may have significance for intrauterine development of the embryo and fetus.

Confined placental mosaicism type I (aneuploidy in cytotrophoblast associated with chromosomally normal placental stroma and fetus) is the commonest one. The aneuploid cell line in the cytotrophoblast detected in prenatal CVS has been shown to persist throughout the entire gestation in the same proportion as seen at 11 weeks gestation. (Kalousek et al, 1988). Although pregnancies with this type of confined chromosomal mosaicism may progress to term uneventfully and result in a normal liveborn infant, some pregnancies with type I mosaicism were described to be associated with unexplained intrauterine fetal death, intrauterine growth retardation and perinatal morbidity (Johnson et al, 1988). Abnormal placental function interfering with fetal development is therefore anticipated in pregnancies with documented placental mosaicism and gestational complications. It needs to be established whether clinically significant trophoblast malfunction is dependent on the presence of specific trisomic clones or certain proportion of trisomic/diploid cells or both.

Chromosomal mosaicism confined only to the chorionic villous stroma represents the type II of mosaicism seen in placenta. It is less common than mosaicism confined to cytotrophoblast and its effect on fetal intrauterine survival is largely unknown. It has been described in both normal pregnancies and pregnancies with fetal IUGR. (Kalousek et al, 1987; Kalousek and Dill, 1983).

Presence of the diploid cell line confined to cytotrophoblast in otherwise non-mosaic aneuploid conceptions, confined placental mosaicism type III, appears to provide in trisomies 13 and 18 a protective effect and facilitate their intrauterine survival (Kalousek et al, 1989). On CVS prenatal diagnosis this type of mosaicism may lead to false negative diagnoses in pregnancies where only direct preparation is analysed (Simoni, 1987).

There are many unanswered questions regarding the significance of confined placental mosaicism in human development. Extensive population cytogenetics data are available for spontaneous abortuses, stillbirth and liveborns, but no comparative data for their placentas. Although it has been shown that confined placental mosaicism, type III, may facilitate intrauterine survival of specific chromosomal trisomies the role of other types is less clearly defined. Confined placental mosaicism has not been previously reported in spontaneously aborted conceptions except for 2 cases involving type II mosaicism with the normal diploid cell line in the cytotrophoblasts and the tetraploid cell line in cultured cells (Yu et al, 1987), which were interpreted as tissue culture artefact. Clearly, the need for more studies of confined placental mosaicism in spontaneous abortuses exists since some studies have indicated high level of mosaicism in cultured tissues from early pregnancy losses (Warburton et al., 1978).

Data from CVS and studies of 2nd and 3rd trimester placentas indicate that in the placenta mitotic errors are common and occur in both chromosomally normal and abnormal conceptions. Their incidence in conceptions at 9-11 weeks of gestation has been estimated at 1-2% (Mikkelsen and Ayme, 1987). Mitotic non-disjunction seen in placentas of chromosomally normal conceptuses appears to involve different chromosomes than meiotic non-disjunction. The most frequently involved chromosomes seen in trisomic clones detected on CVS are #3, #7, #8, #12, and #15. The incidence of mitotic errors in placentas of trisomies 13 and 18 is significantly higher than non-disfunction rate in chromosomally normal conceptions. It equals to the survival of trisomy 13 and 18 conceptuses, which has been calculated to be 5%. The findings in placenta suggest that mitotic errors exist at certain basic constant rate in all dividing tissues and their persistence is dictated by cell selection which may be specific for each tissue. For example, a cell line with trisomy 3 may be viable in tissues of trophoblastic origin but non viable in extraembryonic mesenchyme or embryo proper. A different viability for cells with trisomy 2 may allow their survival and their significant representation in extraembryonic mesenchyme but may prevent their presence in the cytotrophoblast and the embryo proper. It is possible that in addition to different types of confined placental mosaicism there are several yet undiscovered types of confined constitutional mosaicism affecting different tissues in developing embryo/fetus.

Obviously further studies of placentas involving both direct preparations from trophoblast and cultures from villous stroma, chorion and amnion at different stages of intrauterine development are necessary for our understanding of different mutational events taking place during the implantation period and their effect on intrauterine development of both placenta and embryo/fetus. Detailed placental studies at term either confirming the findings of confined placental mosaicism diagnosed by CVS or detecting previously unsuspected mosaicism are specially needed as only in such cases abnormal obstetrical findings such as IUGR or unexplained intrauterine death can be meaningfully associated with confined placental mosaicism.

REFERENCES

Bond DJ, and Chandley CA, (1983). Aneuploidy, Oxford Monographs on Medical Genetics No. 11, Oxford University Press, pp.15-22.

de Grouchy J, and Turleau C, (1982). Atlas des maladies chromosomique. Expansion Scientifique Francaise.

Gardner RL, (1978). The relationship between cell lineage and differentiation in the early mouse embryo. In Gehring AV (ed): Genetic Mosaics and Cell Differentiation, Springer-Verlag, Berlin, Heidelberg, 205-241.

Jacobs P, and Hassold T, (1987). Chromosome abnormalities: Origin and etiology in abortions and livebirths. In Human Genetics. F. Vogel, K. Sperling (Eds), Springer-Verlag. Berlin, Heidelberg, 233-245.

Johnson A, Warpner R, Barr M, Jackson L. Aneuploid/polyploidy confined to cytotrophoblast, IV International Conference on Chorionic Villus Sampling, Athens, May 27-28, 1988.

Kalousek DK, (1985). Mosaicism confined to chorionic tissue in human gestations. In: Fraccaro M, G. Simoni, G. Brambati (eds): First Trimester Fetal Diagnosis. Springer-Verlag, Berlin, Heidelberg, New York, 130-136.

Kalousek DK, Barrett I, McGillivray BC, (1989). Placental mosaicism and intrauterine survival of conceptions with Trisomy 13 and 18. Am J. Hum Genet (in press).

Kalousek DK, Dill FJ. (1983). Chromosomal mosaicism confined to the placenta in human conceptions. Science 221:665-667.

Kalousek DK, Dill FJ, Pantzar T, McGillivray BC, Young SL, Wilson DR. (1987). Confined chorionic mosaicism in prenatal diagnosis. Hum Genet 77:163-167.10.

Kalousek DK, Howard-Peebles PN, Dorfman A, Black SH, Schulman JD, (1988). Confined trophoblastic mosaicism for trisomy 13 in the normal newborn. Amer J. Hum Genet 43:A237.

Markert CL, Petters RM, (1978). Manufactured hexaparental mice show that adults are derived from three embryonic cells. Science 202:56-68.

Martin RH, Mahadevann MM, Taylor PJ, Hildebrand K, Long-Simpson L, Peterson D, Yamamoto J, Fleetham J, (1986). Chromosomal analysis of unfertilized human oocytes. J Reprod Fertil 78:673-682.

Mikkelsen M, and Ayme S, (1987). Chromosomal findings in chorionic villi. A collaborative study. In Human Genetics. Vogel F, Sperling K, (Eds): Springer-Verlag. Berlin, Heidelberg, 589-606.

Pagon RA, Hall JG, Davenport SLH, Aase J, Norwood TH and Hoehn HW (1979). Abnormal skin fibroblast cytogenetics in four dysmorphic patients with normal lymphocyte chromosomes. Am. J. Hum. Genet. 31:54-61.

Peltomaki P, Knuutila S, Ritvanen A, Kaitila I and de la Chapelle A (1987). Pallister-Killian syndrome: Cytogenetic and molecular studies. Clin. Genet. 31:399-405.

Plachot M, Veigu A, Montagnt J, Grouchy J, Calderon G, Lepretre S, Junca A, Santalo J, Elisabeth C, Mandelbaum J, Barri P, Degoy J, Cohen J, Egozcue J, Sabatier JC, Salat-Baroux J, (1988). Are clinical and biological IVF parameters correlated with chromosomal disorders in early life. A multicentric study. Hum Reprod 3:627-639.

Simoni G, Brambati B, Danesino C, Rossela F, Terzoli BL, Ferrari M, Fraccaro M, (1983). Efficient direct chromosome analyses and enzyme determinations from chorionic villi samples in the first trimester of pregnancy. Hum Genet 63:239-257.

Simoni G, Cimelli G, Cuoco C, Terzoli GL, Rossella F, Romitti L, Dalpra L, Nocera G, Tibiletti MG, Tenti P, Fraccaro M, (1985). Discordance between prenatal cytogenetic diagnosis after chorionic villi sampling and chromosomal constitution of the fetus. In: Fraccaro M, Simoni G, Brambati G, (eds): First Trimester Fetal Diagnosis. Springer-Verlag, Berlin, Heidelberg, New York, 137-143.

Simoni G, Fraccaro M, Gimelli G, Maggis F,, Bricarelli DF, (1987). False-positive and false-negative findings on chorionic villus sampling. Prenatal Diag 78:671-672.

Vekemans MJJ, Perry TB, (1986). Cytogenetic analysis of chorionic villi: a technical assessment. Hum Genet 72:307-310.

Warburton, B, Yu, CY, Kline, J and Stein, Z, (1978). Mosaic autosomal trisomy in cultures from spontaneous abortions. Am J Hum Genet 30:609-617.

Watt JL, Templeton AA, Messins I, Bell L, Cunningham P, Duncan RO, (1987). Trisomy 1 in an eight cell human embryo. J Med Genet 24:60-69.

Yu C-Y, Warburton D, Yu MT, (1987). Direct chorionic villus preparations for karyotpying spontaneous abortions. Am J Hum Genet 41:A147.

Molecular and Cytogenetic Studies of
Non-Disjunction, pages 165–181

THE EFFECT OF MATERNAL AGE ON THE FREQUENCY OF TRISOMY: CHANGE IN MEIOSIS OR IN UTERO SELECTION?

Dorothy Warburton

Departments of Pediatrics and Genetics and Development, College of Physicians and Surgeons, Columbia University, New York, New York 10032

INTRODUCTION

The relationship between Down syndrome and maternal age was recognized long before the syndrome was known to result from trisomy for chromosome 21. Although We now have detailed information about many aspects of this relationship,we are still far from understanding the biological basis for the exponential increase in risk of trisomy which occurs in women after the age of 30. In this paper I will briefly review how the available information bears on various hypotheses that have been put forward to explain the effect of maternal age, emphasizing the evidence that comes from our own studies on trisomy in spontaneous abortions.. At the end I will discuss an approach to the question that might be testable in humans.

The four general hypotheses that I will discuss are shown in Table 1. Hypothesis A, the oldest, and still perhaps the most compelling, is that the increasing rate of trisomy with maternal age is due to a maternal age dependent rate of errors at maternal meiosis I (MMI). It should be noted that the intuitive notion that an oocyte which has been arrested in meiotic prophase for many years will become more error prone with time does not really constitute a scientific explanation.

Hypothesis B is what I will call gamete ageing.This was originally proposed by German (1968) and has been extended by others such as Jongbloet (1971) who

TABLE 1. Hypotheses concerning maternal age and trisomy

A	There is an increased rate of error at maternal meiosis I associated with ageing oocytes.
B(1)	In older couples there is a decreased opportunity for conception, due to lower intercourse frequency. This results in an increase in aneuploid gametes, due to misdivision in ageing eggs or selection among ageing sperm in the male tract.
B(2)	Aneuploid gametes are more likely to achieve fertilization in older women.
C	Paternal age also is associated with an increase in trisomy, due either to an increased rate of meiotic error at spermatogenesis, or a selection phenomenon as in B.
D	In older women there is a decreased rate of spontaneous loss of trisomic conceptions before birth (relaxed selection).

proposed that "overripe" oocytes were prone to nondisjunction, and more recently Martin-DeLeon and Williams (1987) who proposed that aneuploidy was increased when sperm were aged in the male tract. These theories (B1) postulate that aneuploidy increases with maternal age because of a decrease in the frequency of intercourse, leading to a longer interval between ovulation and fertilization, or, in the case of the male, between sperm differentiation and ejaculation. Some data from experimental animals support these effects for both eggs and sperm. Ageing of oocytes in the reproductive tract could lead to a meiotic error only in maternal meiosis II (MMII): sperm ageing could act only by changing the frequency with which aneuploid sperm achieve conception, since male meiosis is complete before sperm differentiation. Hook (1983) suggested a variation on this hypothesis (B2),that differences in the environment provided by an older woman may increase the likelihood that a chromosomally aberrant egg or sperm achieves fertilization.

Hypothesis C, which I will address very briefly, is that there are also sources of aneuploidy increased by advancing paternal age. Due to the high correlation of maternal and paternal ages this could result in an apparent maternal age effect in some cases.

Hypothesis D, which I will discuss in more detail, has most recently been proposed by Ayme and Lippman-Hand (1982) and Stein et al.(1986). It has been termed "the relaxed selection" hypothesis by Hook (1983). It differs from all the other hypotheses in that it assumes the frequency of trisomy is the same at conception across all maternal ages. A decrease in the rate of loss of trisomic conceptions during gestation is assumed to lead to an increased frequency of trisomy among livebirths of older women.

THE PARENTAL ORIGIN AND STAGE OF MEIOTIC ERROR

In the 1970's data began to accumulate concerning the parental and meiotic stage of origin of trisomy 21, and also for other human trisomic conditions found mainly among spontaneous abortions. Hypothesis A was supported by the finding that about 70% of cases of Down syndrome, and probably even higher proportions of other trisomies, were due to MMI failure (see data in Hook, 1983). Error at this stage alone could easily account for the entire maternal age relationship. At this symposium we have heard how DNA analysis can define more precisely the events leading to MMI error: possible changes in rates of recombination have received special attention. No one however has speculated on how this might interact with maternal age.

Direct analysis of human oocytes at metaphase I or metaphase II is now possible using the limited numbers available through in vitro fertilization procedures,so that it should be possible to test directly whether there is a higher proportion of aneuploid oocytes in older women, as predicted by hypothesis A. The one study so far which looked at the data stratified by age (Pellester and Sele, 1988) was not able to demonstrate an increase, but the sample size was very small.

Hypothesis A also predicts that trisomies occurring

for reasons other than error at MMI will not show an increased maternal age. Under hypothesis D and some variations of hypothesis B, the same maternal age effect is expected regardless of origin of the parental origin of the trisomy.

TABLE 2. Mean Maternal Age by Origin of Trisomy

	Paternal Origin	Maternal Origin		
		MI	MII	All
Cytogenetic data				
Spontaneous abortions[1]	31.5 (8)	31.9 (60)	37.0 (7)	31.9 (84)
Trisomy 21 livebirths[2]	29.1 (35)	31.1 (79)	31.8 (25)	31.2 (125)
Molecular data				
Trisomy 21[3]	29.3 (7)			31.9 (42)
XXY[4]	27.1 (28)	32.5 (16)	27.4 (4)	31.0 (28)

1.Hassold, this symposium
2.Mikkelsen (personal communication)
3.Buraczynska et al., this symposium
4.Jacobs, this symposium

Table 2 shows the data of Hassold (1989), presented at this symposium, on trisomies found in spontaneous abortions whose origin was determined cytogenetically. Although the number of cases of paternal origin is very small, there is no apparent difference in maternal age between cases due to MMI and cases of other origin. Data from cases of trisomy 21 in livebirths (reviewed in Hook, 1983 and Stein, 1986) are more extensive and have led to the same conclusion.Scandinavian data of this type,collected by Dr. Mikkelsen, are shown in Table 2. It was this finding which led Ayme and Lippman-hand (1981) and Stein (1986) to propose a relaxed selection hypothesis. A maternal age effect for cases of maternal meiosis II error can be explained under hypothesis A if maternal age related differences in the oocyte can lead to error at both MI and MII. However an increased maternal age for paternally derived cases is not consistent with hypothesis A, but is predicted by the relaxed selection hypothesis and is also compatible with hypothesis B.

Another possible explanation for these results is a paternal age related increase in errors of paternal origin, leading to an associated increase in maternal age. There is still some controversy about the effect of paternal age on trisomy 21 frequency, (Stene et al. 1981; Stene, 1988), but most studies have been unable to demonstrate such an effect (Hook, 1982; Hook,1984). Direct evidence against a paternal age effect comes from studies of sperm chromosome complements in men of different ages. The results are inconsistent, in that Martin and Rademaker (1987) found a decrease in aneuploid sperm chromosome complements with increasing paternal age, while Brandriff et al.(1988) found no differences across paternal age groups. However, neither study supports an increase in sperm aneuploidy with paternal age.

Carothers (1987) showed that a small degree of misclassification of parental origin (as low as 8%) would lead to a false impression that maternal age was increased in cases of paternal origin. Many cytogeneticists are willing to admit the possibility of this degree of error when cases are classified on the basis of cytogenetic polymorphisms. It is now possible to determine parental origin by using the more reliable method of DNA markers (Stewart et al. 1988), and the results are beginning to be collected.

These data are also shown in Table 2. For cases of trisomy 21 (data from Buraczynska et al. 1989) there is still not a significantly lower maternal age for cases of paternal origin, although the number of these cases is still very small. At this meeting Jacobs (1989) has also presented data on the maternal age in cases of XXY and XXX trisomy with known origin. Here it appears that only cases due to MMI error have an increased maternal age. Why the sex chromosome anomalies might behave differently from autosomal trisomies in this respect is unclear. The limited pairing and one obligatory chiasma of the X and Y bivalent in males is very different from that for autosomal pairs, and male error is much more frequent than it is for autosomal trisomies. It is also true that the XXY and XXX karyotypes are subject to much less selection in utero, so that relaxed selection with increasing maternal age would be less likely to have any effect.

The data on the XXY anomaly are not strong evidence against the relaxed selection hypothesis, since in utero selection against this trisomy is not nearly as effective as that against other aneuploidies. They do indicate however that maternal meiosis I errors occur more often in older women, independently of any selection,

FETAL LOSS OF TRISOMIES

The data concerning the maternal age relationship of trisomies in fetal death provide strong evidence against the relaxed selection hypothesis.

The general shape of the maternal age curve for liveborn trisomies was addressed in some detail by Hook (1989). Figure 1 shows the same kinds of curves for the trisomic conceptions found in a large series of consecutively studied spontaneous abortions, plotted as a frequency of trisomy among chromosomally abnormal cases.(adapted from Risch et al.1986) For trisomy of most chromosomes there is a significant rise in incidence with maternal age, with a shape similar to that for trisomy 21, and an increase in slope occurring at around age 32. For trisomies 2 and 16 there is also a rise with maternal age, but the slope is constant across maternal age categories, and not as steep. For the largest chromosomes 3-5 there appears to be no increase with maternal age. Data presented by Hassold and Chiu (1985) from an Hawaian study of abortion show a remarkably similar set of curves. Both studies also show that the increase in mean maternal age for cases of trisomy in spontaneous abortions is greatest for the smaller chromosomes, and absent altogether for most larger chromosomes.

Table 3 shows maternal age data by trisomy type from the New York study. When the New York and Hawaian data are considered together there is a significant regression of mean maternal age against chromosome length (Hassold et al.1984). This is not related to whether the chromosomes are acrocentric or contain NOR's, since the non-acrocentric chromosomes show the same maternal age effect as acrocentric chromosomes of the same size. Table 5 also shows that there is no maternal age effect for other chromosome anomalies in spontaneous abortion such

FIGURE 1. Maternal age curves for trisomies in spontaneous abortions (New York data)

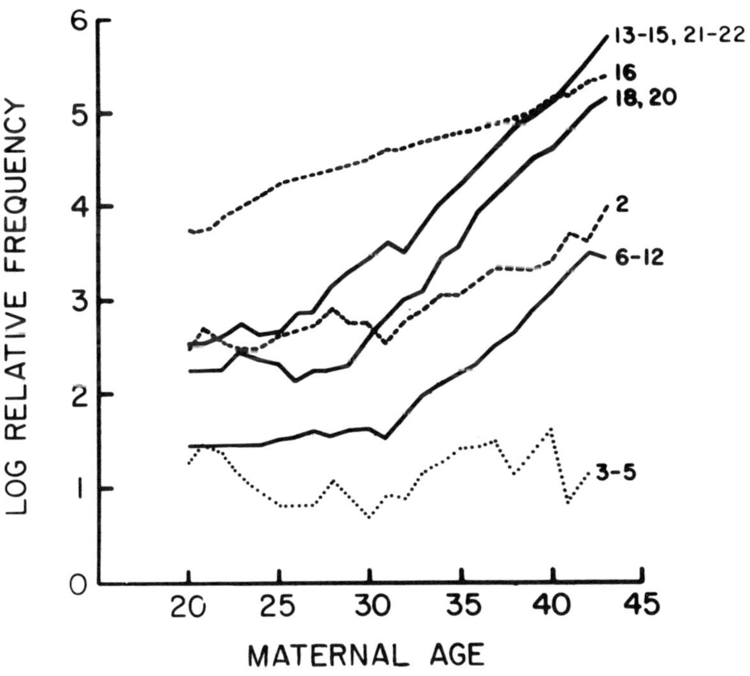

TABLE 3. Mean maternal age for spontaneous abortions with chromosome abnormalities (New York study)

		mat.age ± se	(n)
Trisomies	2-6	26.8 ± 0.9	(65)
	6-12	30.3 ± 0.8	(81)
	13	30.8 ± 0.8	(53)
	14-15	32.6 ± 1.2	(85)
	16	29.4 ± 0.4	(200)
	18	33.6 ± 1.5	(23)
	17,20	33.9 ± 2.1	(22)
	21	33.2 ± 1.0	(53)
	22	33.5 ± 0.8	(55)
Monosomy		24.4 ± 0.2	(206)
Triploidy	XXX	26.3 ± 0.6	(84)
	XXY	26.7 ± 0.6	(97)
Livebirths		26.0 ± 0.5	

as triploidy, and, in fact there is a *decreased* age for abortions with monosomy X as compared with livebirths in the same population.

The difference in maternal age relationship by chromosome size is consistent with the idea of an effect involving pairing relationships at meiosis I, since large chromosomes,with a greater number of chiasmata, might be less likely to separate prematurely. On the other hand it is hard to imagine a relaxed selection process which could discriminate reliably between trisomies on the basis of chromosome size. A maternal age effect is seen even for trisomies which are not capable of surviving to term. In these cases any relaxation of selection must be postulated to move pregnancies normally lost before recognizable pregnancy into the period where they survive to become recognizable spontaneous abortions.

The fact that spontaneous abortions with viable trisomies show an increased maternal age is in itself evidence against the simplest form of the relaxed selection hypothesis, which would predict that younger women would abort trisomies more often, while older women would tend to keep them till birth. We have previously presented (Warburton, 1986) a detailed analysis of the rate of survival of trisomy 21, using maternal age specific rates of spontaneous abortion, trisomy 21 in spontaneous abortions and trisomy 21 in livebirths. With this data one can estimate the proportion of trisomy 21 in recognized conceptions which is eliminated before birth for each maternal age class .As shown in Table 4 there is no decrease in rate of selection against trisomy 21 with maternal age: in fact there is a suggestion of a *decrease* in survival with increasing maternal age.

TABLE 4. Maternal age specific rates of the probability of spontaneous abortion for recognized conceptions with trisomy 21.(from Warburton et al. 1986)

Maternal age	Probability of spontaneous abortion for a conception with trisomy 21
15-19	0.54
20-24	0.55
25-29	0.53
30-34	0.65
35-39	0.52
40-44	0.76
45+	0.84

This is corroborated by looking at the mean length of gestation for viable trisomies which are spontaneously aborted at different maternal ages. The gestation at abortion for trisomies 21 and 18, as well as some other chromosome abnormalities, is actually less in older mothers than in younger mothers, contrary to the relaxed selection hypothesis.(Warburton, 1986) Table 5 shows another way of analyzing the data, where we divide spontaneously aborted pregnancies by whether there was a well-developed fetus when abortion occurred. For trisomy 21 and 18 the pregnancies which achieved a greater degree of development had a *lower* mean maternal age. For trisomy 13 and triploidy there is no difference between the two, and for monosomy there is again a suggestion that the age is lower in the fetuses developing the furthest.

TABLE 5. Degree of embryonic development and maternal age in chromosomally abnormal pregnancies (New York study).

	Mean maternal age (number)			
	Embryo less than 30 mm.		Fetus > 30 mm.	
Trisomy 21	33.7 ± 1.1	(38)	30.3 ± 2.5	(11)
Trisomy 18	35.4 ± 1.7	(14)	30.8 ± 2.7	(9)
Trisomy 13	30.7 ± 0.9	(47)	31.3 ± 3.0	(3)
Triploidy	26.5 ± 0.4	(171)	25.6 ± 2.4	(8)
Monosomy X	25.1 ± 0.4	(188)	21.9 ± 1.3	(9)

These data therefore strongly argue against the idea that selection against trisomic conceptions is less efficient with increasing maternal age. However it is formally possible that *all* the differences in selection occur before recognized pregnancy, in the "black box" between conception and the time a spontaneous abortion can be recognized and karyotyped. Stein et al.(1986) suggested that this might occur at the first mitotic division of the embryo, where an extra chromosome might be more efficiently eliminated in younger women.

Perhaps the strongest evidence against the relaxed selection hypothesis is the fact that there is no maternal age association for unbalanced chromosome anomalies other than trisomy. Triploid and monosomy X conceptions show no increase in frequency with advanced maternal age, nor do unbalanced translocations (Hook, 1987) including unbalanced translocations leading to Down syndrome (Hook, 1983). Thus relaxed selection would have to involve a process that depended upon a maternal recognition of the number of centromeres present rather than any general phenomenon related to survival of an abnormal conceptus with chromosomal imbalance.

THE LIMITED OOCYTE POOL MODEL

Finally I would like to address the mechanism through which an increase in MMI errors might occur. One previous hypothesis, known as "the production line hypothesis" postulated that oocytes matured in the same order in which they were differentiated and entered meiotic prophase. It was then postulated that oocytes entering meiosis later in development might have a different number of chiasmata than those which entered early, and that this was the basis of the difference in the rate of meiosis I nondisjunction.(Henderson and Edwards, 1968).No direct evidence to support this production line hypothesis exists, and some data argue against it (Polani and Jagiello, 1976).

I would like to propose another hypothesis that is compatible with available data. I shall call this the "limited oocyte pool" model. From very early development, a portion of available ovarian follicles begin progressive development, with changes in the granulosa cells, increase in size and fluid content, and changes in hormonal status. At what is known as the antral stage an oocyte is ready to complete meiosis I and achieve ovulation, when stimulated by FSH. In primates usually only one oocyte per cycle completes the final stages of development. As part of this process a feedback mechanism leads to repression of FSH, so that the rest of the antral oocytes cannot proceed with development and degenerate. The method by which the one oocyte is selected for continued development is unknown, but the follicle that will eventually undergo ovulation appears to be the one which emerges from the preantral stage concomitantly with high concentrations of plasma FSH. Under exogenous FSH stimulation other follicles can be stimulated to complete development. There is direct evidence that the number of follicles in the antral stage decreases with increasing maternal age, going from about 7 per ovary in a woman of 25-32 in one study, to about 2 in women 41 to 45 and less than one in women over 45 (Peters and McNatty, 1980). As the number of late stage oocytes decreases it will be less likely that any oocyte is at precisely the optimal stage for FSH stimulation, so that oocytes which are not quite ready or slightly "overripe" may then be selected. If one postulates that such oocytes are more likely to undergo nondisjunction at the first meiotic division, it follows that the ovulated oocytes of older women will have higher rates of aneuploidy. The evidence of Eichenlaub-Ritter et al. (1988) that oocytes in older mice spend a decreased time in metaphase I is consistent with the idea that these oocytes are stimulated to complete meiosis at a different point in maturation than those of younger mice.It is also possible that oocytes which are error prone are less likely to be selected for other reasons. Any mechanism by which selection during maturation tends to favor those eggs which are less likely to be aneuploid will produce an exponential increase in the probability of an aneuploid oocyte as the available pool size decreases.

Brook et al.(1984) proposed a similar model to explain the results of experiments with ovariectomized mice. When one ovary was removed the number of ovulations remained the same, but the available oocyte population was depleted faster. At the same time aneuploidy rates rose faster than expected according to the chronological age of the animal. This supports the idea that it is the depletion of oocytes rather than the chronological age which is necessary for the maternal age related increase in aneuploidy.This is also supported by the fact that those strains of mice where a maternal age related rate of aneuploidy can be demonstrated are those with the shortest reproductive life span whose ovaries eventually become depleted of oocytes. In man there is also some data to suggest that the incidence of trisomy is higher than expected for maternal age in the offspring of Turner mosaics who achieve pregnancy despite a depleted oocyte pool (Reyes et al, 1976; King et al. 1978). An increased risk for trisomy in very young women might also be explained by this theory, since at the first ovulatory cycles the relationship between the FSH stimulus and follicle maturation might be imperfect. The theory is also compatible with data showing that the risk of a subsequent trisomic pregnancy increases for young women who have had a trisomy, but only to about the same rate found in women over 40: at this point the oocyte pool is nearing its lower limit for successful ovulation.

The theory would also predict that exogenous FSH stimulation might lead to a higher than usual incidence be clearly tested in man, and which has provided conflicting results in rodents. Another testable prediction would be that women who have produced trisomic offspring at a young age, or who have had more than one such conception, might have reduced rates of ovulation, accompanied by relative infertility and early menopause.

CONCLUSION

Table 6 summarizes the previous discussion. The hypotheses discussed concerning parental age are listed across the top, and a plus or minus refers to whether the evidence listed is consistent with the hypothesis.The direct evidence against the relaxed selection hypothesis is strong, and the only evidence supporting it is the increased maternal age in cases of paternal origin. This is also the only strong evidence against hypothesis A, which must probably be modified to include errors at meiosis II, at least in autosomes. If new data continue to show an increased maternal age for paternally derived cases,an explanation other than relaxed selection must be sought. Larger numbers of cases with paternal origin defined by molecular markers, and larger studies of retrieved oocytes at different maternal ages should soon settle the matter. Some variation on hypothesis B might then be seriously considered. It is suggested that the decreasing size of the oocyte maturation pool may play a crucial role in the association of increasing maternal age with aneuploidy. This could be tested using both animal models and human data.

TABLE 6. Summary of maternal age data and its consistency with four hypotheses (see TABLE 1)

	A MMI	B gamete ageing	C pat. age	D relaxed selection
1. Most cases of trisomy are due to MMI error	+	−	+	+
2. Autosomal trisomies of paternal origin have increased mat. age	−	+	+	+
3. XXY cases of paternal origin do not have increased mat. age.	+	−	−	+
4. Most other chromosomal defects show no increase with mat. age	+	+	+	−
5. One study of oocytes found no increase in aneuploidy with mat.age	−	−	+	+
6. There is no increase in sperm aneuploidy with age	+	−	−	+
7. Nonviable trisomies also increase with mat. age	+	+	+	−
8. The smaller the chromosome the greater the effect of maternal age	+	−	+	−
9. In trisomy 21 the rate of fetal loss does not decrease with mat. age	+	+	+	−
10. Length of survival of trisomies does not increase with mat age	+	+	+	−

REFERENCES

Ayme S, Lippman-Hand A (1982). Maternal-age effect in aneuploidy: Does altered embryonic selection play a role? Am J Hum Genet 34:558-565.

Brandriff B, Gordon L, Moore II D, Carrano AV (1988). Human sperm cytogenetics: no increase in chromosomal aberrations with age. Abstracts of the Environmental Mutagen Society, 1988.

Brook JD, Gosden RG, Chandley AC (1984). Maternal ageing and aneuploid embryos-evidence from the mouse that biological and not chronological age is the important influence.Hum Genet 66:41-45.

Buraczynska M, Stewart G, Sherman S, Freeman V, Grantham M, Uchida I, Hassold T, Kurnit D (1989). Studying nondisjunction of chromosome 21 with cytognetic markers on the short arm and DNA markers encompassing the long arm. This symposium.

Carothers A (1987). Down syndrome and maternal age: the effect of erroneous assignment of parental origin. Am J Hum Genet 40:147-150.]

Eichenlauber-Ritter U, Chandley AC, Gosden RG (1988). The CBA mouse as a model for age-related aneuploidy in man: studies of oocyte maturation, spindle formation and chromosome alignment during meiosis. Chromosoma 96: 220-226.

German J (1968). Mongolism, delayed fertilization and human sexual behaviour. Nature 217:516-519.

Hassold T, Chiu D (1985). Maternal age specific rates of numerical chromosome abnormalities with special reference to trisomy. Hum Genet 70:11-17.

Hassold T (1989). This symposium.

Hassold T, Warburton D, Kline J, Stein Z (1984). The relationship of maternal age and trisomy among trisomic spontaneous abortions. Am J Hum Genet 36:1349-1356.

Henderson SA, Edwards RG (1968). Chiasma frequency and maternal age in mammals. Nature 218:22-28.

Hook EB (1982). Paternal age and Down's syndrome genotypes diagnosed prenatally: no association in New York State data.Hum Genet 62:167-174.

Hook EB (1983). Down syndrome rates and relaxed selection at older maternal ages. Am J Hum Genet 35:1307-1313.

Hook EB (1984). A search for a paternal-age effect upon cases of 47,+21 in which the extra chromosome is of paternal origin.Am J Hum Genet 36:413-421.Jongbloet PH (1971). Month of birth and gametopathy. Clin Genet. 2:315-330.

Hook EB (1989). This symposium.

Hook EB, Cross PK (1987). Rates of mutant and inherited structural cytogenetic abnormalities detected at amniocentesis: results on about 63,000 fetuses. Ann Hum Genet 51:27-55.

Jacobs PA (1989). This symposium.

Jongbloet PH (1971). Month of birth and gametopathy. Clin Genet 2:315-330.

King CR, Magenis E, Bennet S (1978). Pregnancy and the Turner syndrome. Obstet Gynecol 52:617-624.

Martin RH, Rademaker AW (1987). The effect of age on the frequency of sperm chromosomal abnormalities in normal men Am J Hum Genet 41:484-492.

Mikkelsen M (1989). Personal communication.

Martin-DeLeon PA, Williams, MB (1987). Sexual behaviour and Down syndrome: The biological mechanism. Am J Hum Genet 27:693-700.

Pellester F, Sele B (1988). Assessment of aneuploidy in the human female by using cytogenetics of IVF failures. Am J Hum Genet 42:274-283.

Peters H, McNatty KP (1980)."The Ovary". Berkeley: University of California Press, p.123.

Polani PE, Jagiello GM (1976). Chiasmata, meiotic univalents and age in relationship to aneuploid imbalance in mice. Cytogenet Cell Genet 16:505-529.

Reyes FI, Koh KS,Faiman C (1976). Fertility in women with gonadal dysgenesis. Am J Obstet Gynecol 126:668-670.

Risch N, Stein Z, Kline J, Warburton D (1986). The relationship between maternal age and chromosome size in autosomal trisomy. Am J Hum Genet 39:68-78.

Stein Z, Stein W, Susser M (1986). Attrition of trisomies as a screening device: an explanation for the association of trisomy 21 with maternal age. Lancet ii:944-946.

Stene J,Stene E, Stengel-Rutkowski S, Murken JD (1981). Paternal age and Down's syndrome: Data from prenatal diagnosis (DFG). Hum Genet. 59:119-124

Stewart GD, Hassold TJ, Berg A, Watkins P, Tanzi R and Kurnit DM. 1988. Trisomy 21 (Down syndrome): studying nondisjunction and meiotic recombination by using cytogenetic and molecular polymorphisms that span chromosome 21. Amer. J. Hum. Genet. 42: 227-236.

Warburton D, Kline J, Stein Z, Strobino B (1986). Cytogenetic abnormalities in spontaneous abortions of recognized conceptions.In Porter IH, Hatcher, NA Willey AM (eds): "Perinatal Genetics: Diagnosis and Treatment,"New York: Academic Press, pp 133-148.

Molecular and Cytogenetic Studies of
Non-Disjunction, pages 183–197

GENETIC RISK FACTORS IN HUMAN TRISOMY 21

Margareta Mikkelsen, Hanne Poulsen and Niels Tommerup

Department of Medical Genetics, The J. F. Kennedy Institute, 7, Gl. Landevej, DK-2600 Glostrup, Denmark

INTRODUCTION

Many of the studies involved in the evaluation of genetic or environmental factors in Down syndrome are based on cases of Down syndrome studied cytogenetically in connection with genetic councelling or in connection with confirmation of the diagnosis after referral to a chromosome laboratory. These studies are based on so-called convenience samples. Other studies are based on data from cytogenetic registrees or birth registrees. To avoid pitfalls and ascertainment biases we studied a population of Down syndrome cases and their parents ascertained as complete as possible in a certain period of time.

MATERIAL

Down syndrome cases from Denmark born between 1.1.1980 and 31.12.1985 were recorded and studied cytogenetically. The chromosomes of the parents were, when available, examined likewise for a period of 4 years. For the last 2-year-period only the parents from two areas were studied. For the same period of time data from prenatal diagnosis were available, and as far as possible the chromosomes of the parents of the trisomic fetuses were studied also. Non-disjunction studies were attempted in all families where both parents were available.

A questionnaire was administered blindly before the results of the chromosome study were available. The questionnaire contained questions about pregnancy, birth, health of parents, occupation, exposition to chemicals, isotopes, medical irradiation, places of residence, socioeconomic status, alcohol consumption, smoking, contraception, coffee/tea consumption and certain hobbies. A complete family history was taken. It included parental and grandparental ages, places of birth, abortions, pregnancies, siblings of parents and of the proband. Consanguinity was recorded.

METHODS

The chromosomes of the probands and their families were studied with the routine methods used in our laboratory, e.g. cells were grown in medium 199 with low folate + 5% newborn calf serum. In all cases Q-banding was applied. Nondisjunction-studies applying chromosomal fluorescence heteromorphisms and AgNOR-variants were carried out following the protocol described in a previous paper (Mikkelsen et al., 1980).

The possible role of an interchromosomal effect on the meiotic behaviour of chromosome 21 was studied. A rearrangement in one of the parents was related to the origin of the trisomy 21.

For the cohorte the number and type of fragile sites in the parents were studied likewise. A minimum of 11 Q-banded cells were analysed routinely from each individual. If one cell with a fragile site was observed, 50-100 additional Q-banded cells were analysed. The findings were related to the origin of the extra chromosome 21.

The results of the nondisjunction-study were compared to a number of possible environmental factors and parental ages. Fathers giving the extra chromosome were compared with fathers, who did not contribute the extra chromosome. Similarly, mothers, who gave the extra chromosome, were compared with mothers where nondisjunction was paternal.

RESULTS

The maternal age distribution, 35 years and older and 34 years and younger, of 278 infants born 1980-85 with Down syndrome and 153 fetuses diagnosed prenatally is given in Table 1.

TABLE 1. Denmark 1980-85

Maternal ages	Livebirths No.	D.S. liveborn No.	D.S. fetuses No.	Total No.
- 34	298,292	255	27	282
35 -	21,922	23	126	149
Total	320,214	278	153	431

Karyotypes

The karyotype-distribution of the 278 infants and of 119 fetuses studied 1980-84 is given in Table 2. In 318 cases the karyotypes of both parents were examined. In 244 families nondisjunction studies were possible.

TABLE 2.

Karyotype distribution	Fetuses 1980-1984		Newborns 1980-1985	
	No.	%	No.	%
Trisomy 21	115	96.64	252	90.65
De novo translocations	0	0.00	15	5.40
Inherited translocations	2	1.68	4	1.44
Mosaics	2	1.68	6	2.16
Double trisomies	0	0.00	1	0.36
Total	119	100.00	278	100.00

TABLE 3. De novo translocations, involving chromosome 21

Karyotype	Origin		Age Maternal	Age Paternal
46,XY,-14,+t(14;21)	mat.		23	24
46,XY,-14,+t(14;21)	mat.		23	32
46,XY,-14,+t(14;21)	mat.		27	32
46,XY,-14,+t(14;21)	?		28	26
46,XY,-14,+t(14;21)	mat.		28	32
46,XX,-14,+t(14;21)	mat.		29	33
46,XY,-14,+t(14;21),9qh+	mat.	9qh+	29	24
46,XY,-14,+t(14;21)	mat.		30	33
46,XY,-14,+t(14;21)	mat.		31	31
46,XY,-14,+t(14;21)	pat.		32	33
46,XY,-13,+t(13;21)	?		28	38
46,XY,-13,+t(13;21)	mat.		30	30
46,XX,-21,+t(21;21)	pat.		24	31
46,XY,-21,+t(21;21)	pat.		27	32
46,XY,-21,+t(21;21)	mat.		33	33

Inherited translocations

Karyotype	Age Maternal	Age Paternal
46,XY,-14,+t(14;21)mat.	27	36
46,XX,-14,+t(14;21)mat.	36	34
46,XY,-15,+t(15;21)pat.	20	22
46,XY,-22,+t(21;22)mat.	27	28

TABLE 4. Nondisjunction Studies 1980-85

Type of failure	Newborns	Amniotic fluids	Total
Maternal I	75	23	98
Maternal II	15	10	25
Maternal ?	10	2	12
Total Maternal	100 (78.7%)	35 (87.5%)	135
Paternal I	12	1	13
Paternal II	13	4	17
Paternal ?	2	-	2
Total Paternal	27 (21.30%)	5 (12.50%)	32
Division I	14	5	
Division II	7	6	
Total informative	127 (69.78%)	40 (64.52%)	167
Partiel informat.	21	11	32
Inclusive	34	11	45
Total No. of cases studied with QM and NOR	182	62	244

A de novo translocation was observed in 5.4% of newborns with DS, but non in the fetuses. There was a significant difference reflecting the different maternal age distribution of the two populations.

A comparison of the karyotypes of the newborns to the results found in a study of Greater Copenhagen 1960-72 (Mikkelsen et al., 1976) did not show significant differences between the distribution of trisomy 21 to other aberrations. However, the number of de novo translocations was higher in the period 1980-85. The type of de novo rearrangements and inherited translocations is given in Table 3.

Nondisjunction studies

The results of the nondisjunction-studies for the cases where both parents were available, is given in Table 4. Altogether 244 of such families could be studied. Seventy per cent were informative. The percentage of paternal nondisjunctional events was significantly higher in newborns compared to fetuses. Mean maternal age for mothers of newborns was 28.9 years, for prenatally diagnosed cases mean matenal age was 35.7 years.

The total no of cases studied for chromosomal heteromorphisms was devided into the two maternal age groups 35 years and older and 34 years and younger. Of the informative cases 75% of the young mothers and 70% of the old mothers showed I meiotic failures. The corresponding figures for II meiotic failure were 24% and 30%.

Interchromosomal effect

Reciprocal translocations were observed in 3 cases, and inversions were found in 5 cases. One father carried an extra marker chromosome. In 3 of these 9 cases the rearrangement was observed in the mother, in 6 cases in the father. The type of parental rearrangements is given in Table 5. When inversions of chromosome 9 were excluded, a possible interchromosomal effect appeared. Reciprocal translocations were observed in 3 out of 563 parents, and only in 27 of 20222 consecutively examined newborns (Nielsen et al.). The incidence of 5.3 per 1000 is significantly

TABLE 5. Distribution of nondisjunction among 252 cases with trisomy 21

Maternal age	First meiotic division failure			Second meiotic division failure			Non-informative	Total
	I	MI	PI	II	MII	PII		
34 -	7	23	1	4	6	0	29	70
- 35	14	70	11	7	19	14	47	182
	21	93	12	11	25	14	76	252

(I = first meiotic division), (II = second meiotic division); M = Maternal; P = Paternal.

TABLE 6. Karyotypes and results of nondisjunction studies in families with parental chromosomal rearrangements.

Proband	Father	Age	Mother	Age	Nondisjunction
47,XY,+21 *	46,XY	35	46,XX,t(11;22)	39	MI
47,XX,t(6;17)+21	46,XY,t(6;17)	39	46,XX	34	MI / PII
47,XX,+21	46,XY,t(6;16)	28	46,XX	27	PII
47,XY,+21 *	46,XY/47,XY,+mar	31	46,XX	28	?
47,XY,inv(10),+21	46,XY,inv(10)	38	46,XX	33	MII
47,XX,inv(9),+21, fra(11)	46,XY	23	46,XX,inv(9), fra(11)	24	MII
47,XY,+21	46,XY,inv(9)	36	46,XX	30	?
47,XY,inv(9),+21 *	46,XY	44	46,XX,inv(9)	43	MII
47,XY,inv(9),+21	46,XY,inv(9)	28	46,XX	29	PII

* amnion fluid

M: Maternal; P: Paternal; I: First meiotic division; II: Second meiotic division; ?: Non-informativ

TABLE 7. Nondisjunction and fragile sites

Case	Proband	Father	Age	Mother	Age	Nondisjunction
I	47,XX,+21	46,XY,fra(2)(q13)	41	46,XX	36	MI
II	47,XY,+21	46,XY,fra(16)(q22)	31	46,XX	30	PII
III	47,XX,+21	46,XY	30	46,XX,fra(19)(q13)	20	MII
IV	47,XX,+21,fra11)(q23)	46,XY	23	46,XX,fra(11)(q23)	24	MII
V	47,XX,+21	46,XY,fra(22)(q13)	28	46,XX	25	II
VI	47,XY,+21,fra(2)(q13)	46,XY	39	46,XX	31	?

M: Maternal; P: Paternal; I: first meiotic division; II: second meiotic division;
?: non-informative.

TABLE 8. Sex ratio

Newborn population studies (Mikkelsen et al. 1980) and present study

Type of failure	Males No.	Females No.	Total No.	Sex ratio
Maternal meiosis I	88	83	171	1.06
Maternal meiosis II	25	14	39	1.79
Paternal meiosis I	24	9	33	2.67
Paternal meiosis II	9	9	18	1.00
Total	146	115	261	1.27

different when compared to 1.34 per 1000 newborns. The nondisjunction occurred in meiosis II in 5, possibly 6 of the cases. In one of the families a rare fragile site was observed together with an inversion (Table 6).

Fragile sites

A rare fragile site was identified in one of the parents in 6 of 252 cases giving an incidence of 1:84. Five of the fragile sites were of the folate sensitive type, and one belonged to the AT specific ligand sensitive type (fra(16)(q22) (FRA16B) (Table 7). Five of the fragile sites were found among the 182 couples at maternal age below 35, giving an incidence of 1:73, and one fragile site was found among the 70 couples with a maternal age of 35 and above, giving an incidence of 1:140 (Table 7). In three of the four informative cases nondisjunction had occurred in the parent carrying the fragile site, and in each of these cases two of the three chromosomes 21 were identical indicating that nondisjunction had occurred in meiosis II. Also, in the fourth of the five cases born to young mothers the nondisjunction had taken place in meiosis II, but the origin (maternal/paternal) could not be identified. In the one case with a maternal age above 35 years (36 years) the failure had occurred at maternal meiosis I, and with the father carrying the fragile site.

Sex ratio

The sex ratio and the type of nondisjunction is given in Table 8. The sex ratio is skewed with more males than females, especially in cases were the nondisjunction occurred at paternal meiosis I.

DISCUSSION

We decided to study the whole population of Down syndrome including all newborns and fetuses aborted after amniocentesis to evaluate eventual environmental or genetic risk factors. The so-called convenience sample of Down syndrome may be biased in many ways. Younger families in the desire of another child, and families believing they are at higher risk because of abortions or the birth of another Down syndrome child in the family may get in contact with a genetic center more often than the average Down-syndrome family.

The study gave good population incidence figures for the period 1980-85. There has been a raising incidence from 1980 to 1985, mainly because of a better diagnosis in preterm infants and especially because of a change in maternal age distribution and survival of light for date infants (Nielsen et al., 1987; Mikkelsen, 1988).

The nondisjunction-study showed a similar picture as previously found in Denmark, namely around 20% paternal failures versus 80% maternal failures. The sex ratio was again skewed with more males than females, especially in paternal meiosis I. There is so far no explanation for the skewed sex ratio which is especially found among Down syndrome cases born to young mothers.

Since Hammerton et al. (1965) reported the possible influence of structural rearrangements on the meiotic behaviour of chromosome 21, it has been discussed if such rearrangements in one of the parents can predispose to the birth of a child with Down syndrome. A possible effect was described by Lindenbaum et al., 1985, and by Couzin et al., 1987. At male meiosis Palmer and Hulten (1983) and Laurie et al. (1984) described an increased chiasma frequency and an interchromosomal effect in male carriers of reciprocal translocations. Uchida and Freeman, 1986, however, could not find a support for the occurrence of an interchromosomal parental effect when applying nondisjunction-studies to their cases. We found significantly more reciprocal translocations in the parents compared to consecutively examined newborns in Denmark. However, the population data were in part from the prebanding period, where e.g. the t(11;22) observed in one of the mothers would not have been diagnosed. Unfortunately, our study was only completely informative in six of the nine cases where a rearrangement in one of the parents was observed. The failure had occurred in meiosis II in 5 of these cases. Two of the cases were reciprocal translocations, and in both cases the carrier of the translocation was the source of the extra chromosome in the trisomic infant. For the cases of inversion of chromosome 9 we could not find a higher total risk in our parental population. But in the three cases informative for the origin of the extra chromosome nondisjunction had occurred in the carrier's gametogenesis. More cases with parental rearrangements should be related to the type of nondisjunction that has occurred in the Down syndrome child to give reliable data about the possible interchromosomal effect at meiosis.

Several reports have mentioned the presence of rare heritable fragile sites along with trisomy 21, and it has been discussed whether fragile sites, especially the fragile site on Xq27.3, will predispose to chromosomal nondisjunction. A higher number of XXY Klinefelter patients were observed in males with the fra(X) syndrome (Fryns and Van den Berghe, 1988). A tendency to nondisjunction of the X-chromosome was observed by Brøndum Nielsen et al. (1986) in tissue culture, and recently Watson et al. (1988) reported on a statistical analysis of fragile X sibships showing a significantly higher rate of trisomy 21 than expected. In the cases where we could show the nondisjunction had occurred in the parent with the fragile site, one would expect that also the proband would express the fragile site. This was, however, only the case once (Table 7). However, also in X-linked mental retardation with the fragile X 20% of male carriers do not express the fragile site, and in half of the obligatory female carriers it is not found (Sherman et al., 1985a,b). Again nondisjunction at meiosis II was observed in four of the five informative cases. Second meiotic failure in parents is relatively rare, only found in 28% of the cases. The association between fragile sites and nondisjunction should be studied further in larger series selected without bias and with studies of the origin of the extra chromosome.

The evaluation of possible environmental factors working on parental meiosis is still in progress.

To evaluate the possible influence of imprinting on survival of Down syndrome patients we evaluated the percentage of early infant death in the different groups of meiotic failures. 58 trisomic patients died in the observation period (Nielsen et al., 1987). For maternal meiosis I failure 16 cases died (21.3%). None of the 15 originated by maternal II failures died. For the paternal failures 6 of 27 died (22.2%). These figures are not statistically different.

When we compared the different chromosomal types 23% of the free trisomics, 30% of the translocation trisomics and none of the mosaics (only 6 patient) had died. Larger data would be essential to evaluate a possibel imprinting factor.

REFERENCES

Couzin D A, Watt J L, Stephen G S (1987). Structural rearrangements in the parents of children with primary trisomy 21. J Med Genet 24:280-282.

Fryns J-P, Van den Berghe, H (1988). The concurrence of Klinefelter syndrome and fragile X syndrome. Am J Med Genet 30:109-113.

Hamerton J L, Giannelli F, Polani P E (1965). Cytogenetics of Down's syndrome (mongolism) I. Data on a consecutive series of patients referred for genetic counselling and diagnosis. Cytogenetics 4:171-185.

Lindenbaum R H, Hultén M, McDermott A, Seabright M (1985). The prevalence of translocations in parents of children with regular trisomy 21: a possible interchromosomal effect? J Med Genet 22:24-28.

Laurie D A, Palmer R W, Hultén M A (1984). Studies on chiasma frequency and distribution in two fertile men carrying reciprocal translocations; one with a t(9;10) karyotype and one with a t(Y:10) karyotype. Hum Genet 68:235-247.

Mikkelsen M (1988). The incidence of Down's syndrome and progress towards its reduction. Phil Trans R Soc Lond B 319:315-324.

Mikkelsen M, Fischer G, Stene J, Stene E, Petersen E (1976). Incidence study of Down's syndrome in Copenhagen, 1960-1971: with chromosome investigation. Ann Hum Genet, Lond 40:177-182.

Mikkelsen M, Poulsen H, Grinsted J, Lange A (1980). Non-disjunction in trisomy 21: study of chromosomal heteromorphisms in 110 families. Ann Hum Genet, Lond 44:17-28.

Nielsen K B (1986). Sex chromosome aneuploidy in fragile X carriers. Am J Med Genet 23:537-544.

Nielsen J, Hansen K B, Sillesen I, Videbech P (1982). Chromosome abnormalities in newborn children. Aetiological aspects. Hereditas 96:109-117.

Nielsen K G, Pilgaard B, Mikkelsen M (1987). Prævalensen og overlevelsen af nyfødte med Down's syndrom i Danmark 1980-1985. Ugeskr Læger 149/32:2170-2173.

Palmer R W, Hultén M A (1983). Chiasma-derived genetic lengths and recombination fractions: a reciprocal translocation 46,XY,t(1:22)(q32;q13). Ann Hum Genet 47:299-310.

Sherman S L, Jacobs P A, Morton N E, Froster-Iskenius U, Howard-Peebles P N, Nielsen K B, Partington M W, Sutherland G R, Turner G, Watson M (1985a). Further segregation analysis of the fragile X syndrome with special reference to transmitting males. Hum Genet 69:289-299.

Sherman S L, Jacobs P A, Morton N E, Froster-Iskenius U, Howard-Peebles P N, Nielsen K B, Partington M W, Suther land G R, Turner G, Watson M (1985b). Further segregation analysis of the fragile X syndrome with special reference to transmitting males. Hum Genet 71:184-186.

Uchida I A, Freeman V C P (1986). Trisomy 21 Down syndrome II. Structural chromosome rearrangements in the parents. Hum Genet 72:118-122.

Watson M S, Breg W R, Pauls D, Brown W T, Carroll A J, Howard-Peebles P N, Meryash D, Shapiro L R (1988). Aneuploidy and the fragile X syndrome. Am J Med Genet 30: 115-121.

Molecular and Cytogenetic Studies of Non-Disjunction, pages 199–220

THE SPINDLE OF MAMMALIAN OOCYTES: ALTERATIONS IN MORPHOLOGY AND PROGRESSION THROUGH MEIOSIS WITH MATERNAL AGE

Ursula Eichenlaub-Ritter

Department of Biology, Gentechnology, University of Bielefeld, POB 8640, 4800 Bielefeld, FRG

INTRODUCTION

The rise in aneuploid zygotes with increasing maternal age, like the high incidence of trisomy 21 in offspring of the aged human female, have been related to a number of events and mechanisms, e.g. disturbed cellular development in the embryo (Henderson and Edwards, 1968), hormonal inbalance in the aged organism, and damage of the oocyte caused by environmental factors or by ageing processes in the cell *per se* during its long retention in the reproductive system before ovulation (reviewed in Bond and Chandley, 1983).

Only in the last few years has the application of new cytogenetic and molecular techniques (e.g. Jacobs et al., 1988; Hassold et al., this volume) combined with early screening and detection of chromosomally inbalanced embryos (e.g. Mikkelson et al., this volume) made it possible to trace the trisomic condition, for example, to either parental gamete, and to provide information on the origin of the aneuploid state. It would appear that most age-related chromosomal aberrations seem to be caused by a non-disjunction event during the first meiotic division of the maturing oocyte (e.g. Hassold and Jacobs, 1984; Jacobs et al., 1988).

It is conceivable that predisposition to aneuploidy results from an accumulation of defects in these cellular components which mediate proper chromosome orientation, separation, or migration to a spindle pole, in those oocytes maturing and ovulated near the end of the reproductive span in the female.

In the past, observations on the, undoubtedly, instrumental basis or organelle translocation and chromosome migration, the microtubules and the spindle apparatus in the oocyte, have been restricted (e.g. Szöllösi et al., 1986). Due to the large size of the mammalian oocyte, systematic screening of cells for age-dependent alterations in their cytoskeleton has not been possible. By advent of immunofluorescence techniques and the availability of tubulin antibodies (e.g. Kilmartin et al., 1982), together with the development of appropriate methods for extraction and flattening of cells (e.g. Balcson and Schatten, 1983; Eichenlaub-Ritter et al., 1986), it has become possible to examine the cytoskeleton of vast numbers of oocytes in some detail (e.g. Wassarman and Fujiwara, 1978; Schatten et al., 1985; Maro et al., 1985). In this communication we present data which correlate morphological and developmental alterations in the spindle and chromosomes of mammalian oocytes, in particular, those of the human, with the age of the female (Eichenlaub-Ritter et al., 1986; Eichenlaub-Ritter et al., 1988).

Since human oocytes are not readily available for immunofluorescent studies the CBA mouse has been used as a model system in most experiments because it also displays a maternal age-related rise in aneuploidy although this is much less pronounced than in the human (Gosden, 1973; Martin et al., 1986, Fabricant and Schneider, 1978). Observations in mice may therefore not be generalized but can still give indications of the mechanisms responsible for predisposition to non-disjunction in humans. For instance, characteristic changes in microtubule and chromosome organization during postovulatory ageing of the oocyte are quite similar in both organisms (Eichenlaub-Ritter et al., 1986;-1988).

In this communication the results of experiments testing for differences in spindle structure and function, chromosome positioning and configurations, timing of events, during maturation and sensitivity of oocytes towards environmental perturbancies with regard to age are presented. They show that the application of standard cytogenetic methods combined with immunofluorescence techniques and examination of spindle structure and function may open up a new perspective to study the aetiology of non-disjunction.

RESULTS

I. DETECTION OF AGE-RELATED ALTERATIONS IN THE SPINDLE

Studies on Spontaneously Ovulated, Metaphase II Arrested Mouse Oocytes

Anti-tubulin immunofluorescence and Dapi staining of spontaneously ovulated oocytes of the CBA mouse after post-ovulatory ageing in vivo revealed distinct changes in the organization of the spindle apparatus and chromosome positioning in the metaphase II arrested oocytes (Eichenlaub-Ritter et al., 1986). Whereas in freshly ovulated oocytes microtubules reside predominantly in the barrel-shaped spindle aligned between the flat poles and the chromosomes at the equator, ageing results in i) assembly of microtubules at focal organizing centers all over the cell, ii) polymerization of astral spindle fibres extending away from the poles towards the cell periphery, and iii) loss of tubules in the spindle. Apart from a rotation of the spindle in aged oocytes which has previously been noticed (e.g. Szöllösi, 1975), the pole-to-pole distance of spindles decreases significantly ($p<0.001$) during ageing from an average of 26.4 µm ($n=48$) to 21.7 µm ($n=38$). Concomitantly, chromosomes which are located at the spindle equator in most freshly ovulated oocytes (89%, $n=82$) become spread and more irregularly aligned off-center of the spindle (61.4% of aged cells; $n=44$). Subsequently, microtubules located throughout the entire cytoplasm become depolymerized gradually. The spindle looses its bipolarity. Chromosomes are scattered over the remnants of the spindle apparatus before a restitution nucleus is formed (Eichenlaub-Ritter et al., 1986). Degeneration of the spindle and less ordered chromosome positioning in postovulatory aged but still fertilizable oocytes could predispose the cell to non-disjunction, aneuploidy, and contribute to the observed reduction in developmental capacity of the postovulatory aged oocytes (Gianfortoni and Gulyas, 1985). However, unlike former hypothesis predicted (German, 1968), these changes in the oocyte are not related to maternal age-dependent aneuploidy since it is in the first and not the second meiotic division that most chromosomes undergo malsegregation (Hassold et al., 1984).

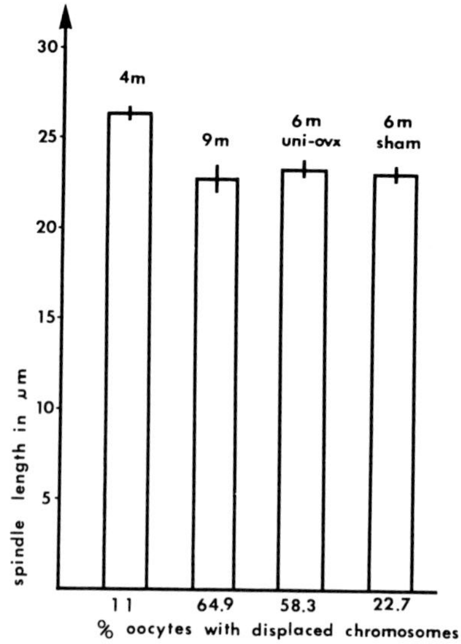

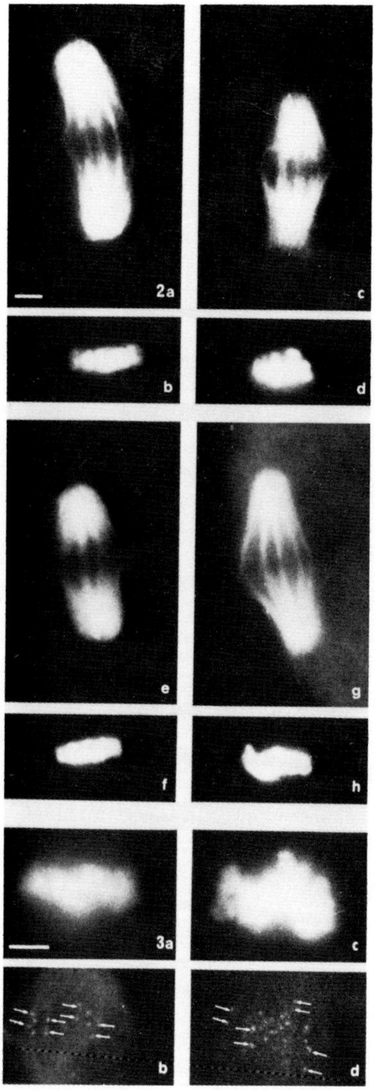

Fig. 1: Spindle length and chromosome displacement in oocytes from young (4 month) aged (9 month), 6 month old unilaterally ovariectomized (uni-ovx) and 6 month old sham control CBA mice.

Fig. 2-3: Tubulin immunofluorescence (2a,c,e,g) and chromosomes (2b,d,f,h) in oocytes from young /2a,b), aged (2c,d) 6 month sham controls (2e,f), and 6 month uni-ovx (2g,h) mice. Some oocytes possess highly ordered chromosomes (3a) with centromeres all aligned at the equator (3b: anti-centromere immunofluoresscence) while predominantly those oocytes of aged mice have chromosomes (3c) and centromeres (3d) which are spread around the equator. Bar: 10μm.

With regard to possible differences in the spindle apparatus with increasing maternal age, or rather a correlation to the end of the female reproductive span, we compared microtubules and chromosome organization of freshly ovulated oocytes from young (2-4 month), aged (9 month and older), and moderately aged (6-7 month) unilaterally ovariectomized (uni ovx) CBA mice. Ovulation ceases much earlier in mice which are unilaterally ovariectomized at an early age (6-8 weeks) compared to females with two intact ovaries. This provides the opportunity to distinguish between the effects of a) the chronological age of the female, and b) the physiological age of the ovary on alterations in cell organization and in the incidence of aneuploidy (e.g. Brook et al. , 1983). While the general morphology of the spindle and distribution of microtubules was similar in all experimental groups (Fig. 2, 3), pole-to-pole distance of spindles varied (Fig. 1). In oocytes of aged, uni-ovx and 6 month old sham-operated control mice the metaphase II spindle was significantly shorter (22.7 µm, 23.3 µm, and 23 µm, respectively; $p<0.001$) than in oocytes of young females (26.4 µm). Thus, changes in spindle length but not the overall organization of microtubules seem to be related to maternal age.

However, chromosome disorder and displacement was most pronounced in oocytes of aged and moderately aged uni-ovx CBA mice (64.9% and 58.3%). In the young as well as the moderately aged sham controls it was considerably lower (11% and 22.7%). If displacement from the spindle equator reflects a preceeding non-disjunction event or indicates that a chromosome has a reduced capacity to attach to microtubules and thus is at a high risk for non-disjunction, these data imply that it is primarily the physiological age of the ovary and its depletion of oocytes which predispose the female to aneuploidy. The age of the mother may be an additional risk factor (Eichenlaub-Ritter et al., 1988). In accordance, cytogenetic analysis in another study using uni-ovx CBA mice revealed that elevated levels of aneuploidy are predominantly correlated with the nearness to the end of the reproductive period of the female and not maternal age *per se* (Brook et al., 1983).

Microtubule Organization, Spindles and Chromosomes in Aged Oocytes of the Human

To detect age-related structural alterations in human oocytes which may lead to aneuploidy or give information on the possible origin of infertility and non-disjunction, unfertilized human oocytes from an IVF programme were examined by indirect immunofluorescence (Eichenlaub-Ritter et al., 1988). Since ethical considerations prevented us from studying any freshly obtained cell, the cytoskeleton of unfertilized oocytes age in vitro for 2, 3 and 4 days was examined. Although, so far, only a limited number or oocytes have been available, a gradual redistribution of microtubules and changes in chromosome configuration were monitored, and some disturbancies related to maternal age. Of 38 unfertilized oocytes available at that time, 30 cells from 16 donors with an average age of 34.5 years could successfully be attached to slides and processed for immunofluorescence.

The spindle of human oocytes aged for 2 days in culture is already rather small; in some cells it is bipolar but often also multipolar (Fig. 4a-d). In contrast to mouse oocytes no cytoplasmic microtubules can be observed, and chromosomes are scattered all over the degenerating spindle in all cells, even those still possessing a bipolar spindle apparatus (fig. 4b). This implies that human oocytes are exposed to rapid degeneration during postovulatory ageing. After two days of ageing they no longer seem fit to develop into a chromosomally balanced, normal embryo.

If oocytes are subsequently aged for one or two more days, distintegration of the spindle proceeds. Microtubules start to radiate out in an astral fashion from the decondensing chromosomes (fig. 4e,f; 3 days) or from the restitution nucleus towards the cell periphery (fig. 4g-i; 4 days of ageing). A similar alignment of tubules has been observed in an oocyte which was probably spontaneously activated and underwent immediate cleavage. Therefore, characteristic alterations in the spindle and chromosomes of ageing human and murine oocytes seem to be very similar, in that in both species chromosome behaviour and the rearrangement of the cytoskeleton mimic to a certain extent those rapid changes taking place in the egg after fertilization (Schatten et al., 1985; Eichenlaub-Ritter et al., 1986; - 1988). A spindle associated polymerization of microtubules appears to be replaced by an interphase-like, cytoplasm-orientated assembly of tubules.

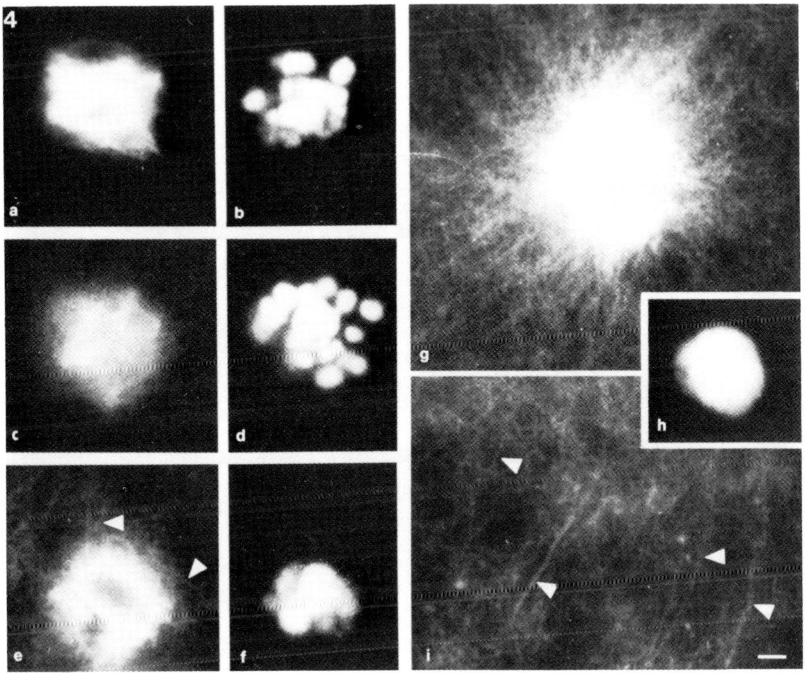

Fig. 4: Antitubulin immunofluorescence (4a,c,e,g,i) and chromosomes (4b,d,f,h) of unfertilized human oocytes aged *in vitro* for 2 days (4a-d), 3 days (4e,f) and 4 days (4g-i). Bar: 5µm.

Apart from the activated oocyte 4 other aberrant oocytes were observed among the cells obtained after IVF (this excludes 6 atretic or damaged oocytes possessing no microtubules and only condensed chromatin, and one GV stage). Two of the unusual cells had two sets of chromosomes and two spindles. In one of these apparently "binucleate", undivided

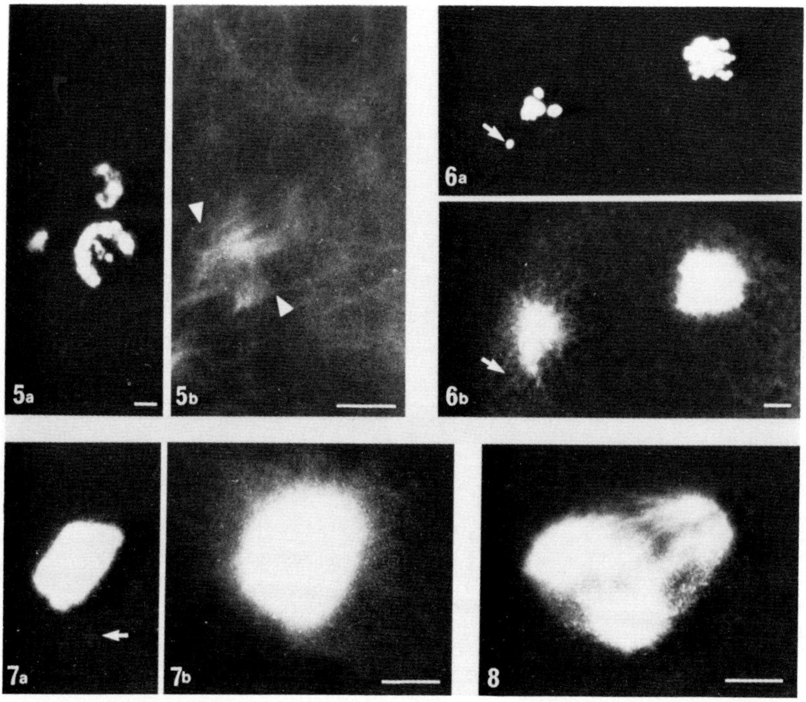

Fig. 5-8: Antitubulin immunofluorescence (5b,6b,7b,8) and chromosomes (5a,6a,7a) in aberrant human oocytes. For explanation see text. Bar: 10μm.

oocytes which had aged for only two days, a tripolar spindle was found (fig. 8). The other cell exhibited in addition to two degenerating spindles, displacement of a chromosome from the rest of the chromatin, and, presumably, unequal numbers of chromosomes in the two spindles (fig. 6a,b). Two more oocytes had only one cluster of chromosomes but as well an individual chromosome lying on its own in the cytoplasm (fig. 7a,b). Such displacement of chromosomes was, in contrast, never found in murine oocytes, not even after ageing. Therefore we assume that these chromosomes were already located outside of the spindle in the freshly ovulated cells, and, because of this, predisposed to malsegregation.

All aberrant oocytes observed so far, apart from the activated cell, came from women 35 years of age or older. Thus, in our sample 28.5% of all oocytes from aged women possessed structural abnormalties. Even if one takes into account the fact that we were looking only at a very small sample population, and a preselected set of cells, i.e. those remaining unfertilized in IVF, these data emphasize the higher risk of structural disturbancies in oocytes of aged females. Although recent karyotyping of unfertilized oocytes from IVF gave contradictory data (Wramsby and Fredga, 1987; Pellestor and Sele, 1988; Placot et al., 1988), and some imply that chromosomally unbalanced oocytes are not particularly frequent in oocytes of the aged female (Pelestor and Sele, 1988), structural aberrations revealed by immunofluorescence cannot be detected by cytogenetics. However, disturbancies of this kind, in particular, e.g. of microtubule or chromosome positioning, may contribute to the much higher frequency of foetal loss and chromosomally unbalanced embryos detected by prenatal diagnosis and studies on aborted and liveborn offspring (see Mikkelson et al., this volume). Therefore, immunocytochemical techniques may be a new tool useful in the detection and analysis of structural disturbancies which lead to age-related rises in aneuploidy.

II) MECHANISMS RESPONSIBLE FOR NON-DISJUNCTION

First Meiotic Division in Mouse Oocytes

Since it is at the first meiotic division that most errors in chromosome separation occur, we examined *in vitro* maturing oocytes to detect spindle malfunction. To be able to compare cells from young and aged females at the same stage of development/maturation, oocytes from hormonally untreated, diestrous, young and old CBA mice were isolated. After rapid removal of all follicle cells, they were placed into culture dishes and induced to undergo spontaneous resumption of meiosis. Germinal vesicle breakdown (GVBD) and polar body formation (PBF) were recorded in the spontaneously maturing oocyte populations, as well as samples processed for indirect immunofluorescence at appropriate intervals. Metaphase II cells can be spread and cytogenetically analyzed.

Immunofluorescence shows that the successive stages of meiosis can be characterized by the typical organization of microtubules and chromosomes in the oocyte. In prophase, after GVBD, unordered microtubules are found in abundance, predominantly in association with the still clustered, condensed bivalents. In prometaphase a bipolar spindle is formed but bivalents are still scattered around the spindle equator, some moving towards, or close to one or the other spindle pole (fig. 9a,b). In metaphase all bivalents are aligned in the center of the spindle, midway between opposite spindle poles (fig. 9c-e). Disjoining of homologes at anaphase of the first meiotic division appears to be slightly asynchronous, so that in a fixed anaphase cell some chromosomes may have already separated while others are located at the equator or are still in the process of migration (fig. 9f,g). In telophase kinetochore microtubules

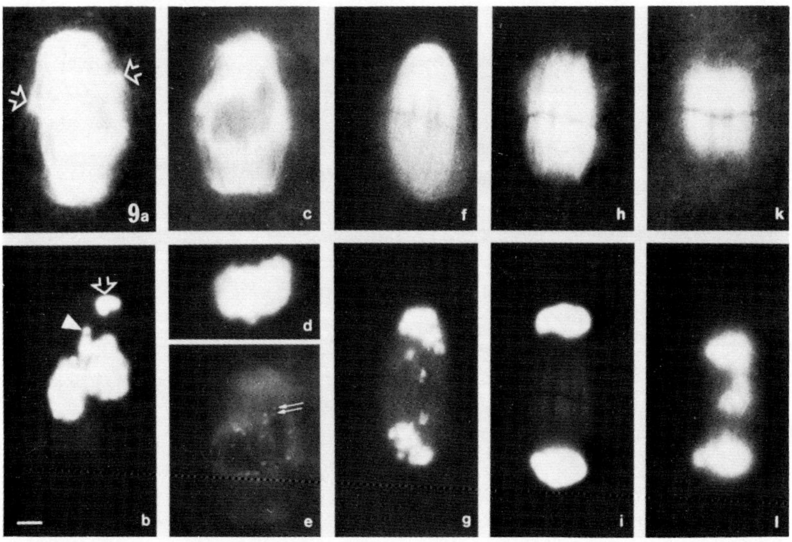

Fig. 9: Microtubules (9a,c,f,h,k), chromosomes (9b,d,g,i,l) and centromeres (9e) of in vitro maturing murine oocytes at different stages of meiosis. Bar: 10µm.

disassemble, the spindle elongates, and separated chromosomes in the presumptive PB or oocyte become clustered and slightly decondensed (fig. 9h,i). After this, the interpolar spindle region, the midbody, becomes constricted and cytokinesis takes place. After a brief prophase II, chromosomes in the oocyte again become aligned at the spindle equator in metaphase II.

Analysis of the changes in the cytoskeleton during maturation demonstrate that, in contrary to ovulated oocytes some parameter e.g. spindle length, cannot be compared between groups of oocytes because they have already undergone alterations in individual cells during meiosis. Also, anaphase lagging cannot unambiguously be detected because of the asynchronous manner of separation of homologes in all cells (fig. 9k,l). Only those chromosomes not migrating at all might be detected in the midbody of telophase spindles. However this type of lagging has not been observed, either in the oocytes of young or of old mice. Also other structural differences in the spindles were not noticed. Disturbancies such as the separation of a single chromosome from the spindle body can be induced by drug-treatment but they have not been examined in untreated cells. General spindle morphology seems to be very similar in the oocytes of young and aged mice. Therefore, if structural aberrations occur in maturing oocytes of aged females, they must be too subtle to be detected by immunofluorescence.

However, in spite of these discouraging findings comparisons of the percentage of oocytes in the different stages of meiosis and observation of GVBD and PBF in oocytes of young and aged mice reveal that there is a distinct difference in progression through maturation in the oocytes of 2-4 month compared to 9-10 month old mice. While GVBD takes place at about the same time or even slightly later in the old age group, PBF in the same cells is triggered much earlier. Concomitantly, we found that 92% of all oocytes of young mice in a representative sample were still in pro-/prometaphase and only 8% in metaphase after 7hrs of maturation (n=38), while a mere 60% of the aged group were in prophase and 36% already in metaphase (n=25); 4% had even started anaphase. Similar observations were made after 8 and 10hrs of culture: a larger number of oocytes from old animals were always in advanced stages of meiosis compared to oocytes from 2-4 month old mice (table 1). Chromosome

TABLE 1:

PERCENTAGE OF OOCYTES OF YOUNG AND AGED MICE IN DIFFERENT STAGES OF MEIOSIS AFTER 7HRS AND 10HRS OF MATURATION IN VITRO:

	7HRS		10HRS	
	YOUNG	AGED	YOUNG	AGED
PRO- TO PROMETAPHASE	92	60	36	12
METAPHASE	8	36	17	17
ANA- TO METAPHASE II	-	4	47	71
N =	38	25	36	35

analysis in metaphase arrested oocytes matured in vitro demonstrates that, at the same time, hyperploid oocytes are more frequent among the cells from the aged compared to the young mice. As such, only 1.6% of all oocytes from young mice (n=63) had more than 20 chromosomes (3.2% had 40 chromosomes) while 5.4% of those from aged mice (n=55) were hyperploid (plus 12.7% with 40 chromosomes). In summary, this suggests that there is a correlation between timing of anaphase trigger/cytokinesis, and incidence of aneuploidy (Boll and Eichenlaub-Ritter, 1988). Since oocytes were isolated at the same stage of the estrous cycle from untreated mice, and cultured and processed under identical conditions we conclude that both, predisposition to aneuploidy and altered development, are intrinsic to the oocytes of the aged females. Since it is during prometaphase that bivalents migrate, become attached and organized on the spindle and oriented towards opposite poles, a shortening of this critical period could be responsible for the frequent malsegregation of a chromosome and predisposition of the oocyte of an aged mammal to aneuploidy.

Sensitivity of Oocytes towards Microtubule-Disrupting Drugs

Earlier studies have shown that the microtubule-disrupting drug colchicine when applied to young and aged mice at the time of oocyte maturation has different effects on the chromosomal constitution of the ovulated oocytes (Tease and Fisher, 1986). Among other factors, maternal age-related differences in the sensitivity of the oocyte spindle towards the drug were held responsible for the altered response. We designed an experimental set up in which i) only the oocyte itself is exposed to the drug, and ii) the block is applied for a known and limited time during maturation. For our experiments we used the microtubule-disrupting drug nocodazole instead of colchicine which has the same binding site on tubulin as the latter. It also leads to disassembly of microtubules but can enter the cell much more rapidly than colchicine and is easily washed out (e.g. Hoebke, et al., 1976). Oocytes were isolated from young and old, diestrous CBA mice which had not been hormonally pretreated and were exposed to the drug *in vitro*.

Immunofluorescence showed that incubation for one hour at low drug concentrations (10 μM nocodazole) results in com-

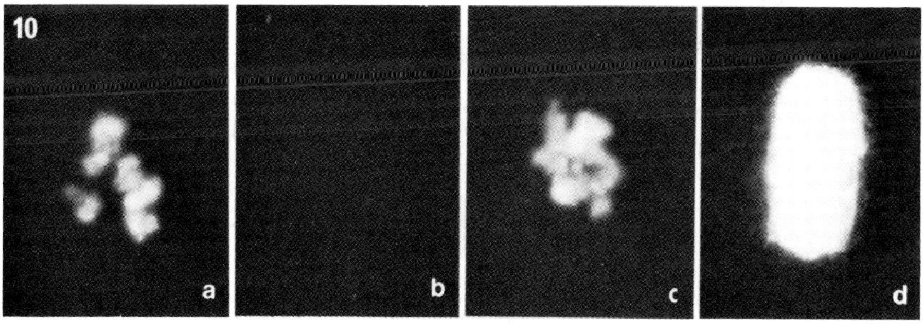

Fig. 10: Chromosomes (10a,c) and antitubulin immunofluorescence in murine oocytes after nocodazole treatment (10a,b) and recovery for 1h (10c,d). All microtubules become depolymerized by the drug but can be reassembled upon recovery.

lete disassemby of all spindle microtubules of prophase cells (up to metaphase), and leads to dispersal of chromosomes in a large number of oocytes (fig. 10a,b). Upon recovery in a drug-free medium, all cells rapidly reorganize a spindle, so that pro- (fig. 10c,d) and even a few anaphase stages are found after an hours recovery period.

The response to nocodazole, in terms of microtubule depolymerization and rapid reconstitution of a spindle is the same in oocytes from both age groups. Interestingly, not only reassembly of a spindle but also progression through meiosis after drug-exposure are similar in oocytes from young and aged mice. For instance, 2.7% oocytes of young and 4% of aged mice had entered anaphase 1hr after recovery from a nocodazole-block (8hrs maturation, 1hr nocodazole, 1hr recovery = 10hrs of culture in total). In the same sample, 32.4% (young) and 24.5% (aged) of oocytes had reached first metaphase. After two hours of recovery (11hrs from the start of spontaneous resumption of meiosis) most cells of both groups had triggered anaphase (table 2).

TABLE 2

PERCENTAGE OF OOCYTES IN DIFFERENT STAGES OF MEIOSIS AFTER A 1HR NOCODAZOLE-BLOCK AND RECOVERY FOR 1HR:

	YOUNG(2-4MONTH)	AGED(9-10MONTH)
PRO- TO PROMETAPHASE	64.9	72
METAPHASE	32.4	24
ANA- TO METAPHASE II	2.7	4
N=	37	25

From this it can be concluded that i) nocodazole treatment causes chromosomes of oocytes to arrest in metaphase. ii) Its action is reversible and a spindle is reformed. iii) Reassembly of the spindle and division is rapid in recovering oocytes, and iiii) further proliferation follows a similar pattern in both age groups.

As expected, nocodazole treatment leads to obvious disturbancies of spindle and chromosome organization in some cells. For example, telophases with a tripolar spindle, or chromosome lagging were observed among the recovered oocytes. Some cells exhibit single chromosomes outside of the spindle apparatus and free of microtubular fibres. Still, obvious structural aberrations are not so frequent, and there is no particular kind of structural aberration restricted to only one of the age groups or absent from one experimental group.

In contrast, cytogenetic analysis of nocodazole-exposed oocytes in metaphase II reveals that non-disjunction and aneuploidy (hyperploidy) are greatly increased by drug treatment (table 3). While absolute rates of aneuploidy

TABLE 3:

PERCENTAGE OF HYPERPLOID OOCYTES OF YOUNG AND AGED MICE AFTER A 1HR NOCODAZOLE-BLOCK, 7HRS AND 8HRS AFTER MATURATION IN VITRO:

	YOUNG (2-4MONTH)	AGED(9-10MONTH)
CONTROL	1.6	5.4
NOCODAZOLE, 7HRS	11.6	10
NOCODAZOLE, 8HRS	10	12.7
N=	63/43/70	55/80/63

(calculated as 2xhyperploidy) are similar in oocytes of young and aged mice (23.2% versus 20%), the increases in hyperploid cells are significant and much higher in the young females (mean 10%) but are more moderate in oocytes of aged animals (mean 4.6%). Thus, there is a difference in the response of the cells to drug-treatment, but, in contrary to expectation, oocytes of young mice seem more susceptible to drug-induced non-disjunction.

To exclude that the difference in response is due to the intrinsic difference in progression through maturation and anaphase trigger, the nocodazole block has also been applied later (instead of 7hrs, 8hrs culture before drug addition). Again, hyperploidy increases compared to controls, but the absolute rate is similar in both groups of oocytes of young and aged mice (table 3). Although the absolute rate of aneuploidy is the same, the difference in the relative increase in hyperploidy remains when the percentage of aneuploidy in the oocytes of young mice treated after 8hrs (10% hyperploids, n=70; relative increase: 8.4%) is compared to that of oocytes from aged females treated an hour earlier (7hrs) (10% hyperploids, n=80; relative increase only: 4.4%). Therefore, the different response of oocytes from both groups cannot result from the altered development *per se*.

Does this mean that oocytes from young mice are more sensitive to the drug and have a more fragile spindle (see Tease and Fisher, 1986)? Some considerations lead to the rejection of this hypothesis. As stressed above, the shortening of the interval from pro- to metaphase in maturing oocytes from aged females may be responsible for the less stringent control in chromosome distribution. Nocodazole disrupts the spindle apparatus in the oocytes of both, young and aged mice, and arrests them in a metaphase stage. In both groups of cells reassembly of the spindle and division take place in a very short period of time (1-2hrs). Therefore nocodazole could abolish the "advantage" of the oocytes from young animals for longer and more controlled orientation of homologes. Since the drug affects both groups of cells in much the same fashion, inherent differences in timing of maturation seem to be abolished and could therefore result in a similar predisposition to non-disjunction/malorientation in oocytes of young and aged mice. Therefore our data rather support than refute the interpretation that it is the disturbancies in timing during maturation and not a damaged or more fragile spindle or a greater number of univalents etc. which is responsible for the age-related rise in aneuploidy in mammals.

CONCULISIONS AND PERSPECTIVES

The presented data show that application of immunofluorescent techniques to the examination of the spindle apparatus and the detection of structural alterations in mammalian oocytes may become a useful tool to trace predisposition to aneuploidy.

For instance, from studies of postovulatory ageing in mouse as well as human oocytes it has become clear that ageing goes hand in hand with distinct alterations of spindle and chromosome organization. Gradual degeneration and loss of temporal and spatial control of tubulin polymerization causing chromosome displacement could, indeed, contribute to the reduced developmental capacity of the aged female gamete (Eichenlaub-Ritter et al., 1986). Since these age-related disturbancies are especially pronounced in the human, they may reflect the overall sensitive state of this cell. In particular, disturbancies in the structural organization of the oocyte could be responsible for the high amount of fetal wastage and chromosomal disorders in the human, which is comparable to no other species (e.g. Hassold and Jacobs, 1984).

In fact, structural aberrations such as the presence of more than one spindle and gross chromosome dislocation appear to be more common in the oocytes of aged rather than of young women. Although the data are preliminary and have to be considered with care, they do show that certain kinds of disturbancies which may not be detectable by conventional cytogenetics can become visible with the present approach. Therefore, immunofluorescence studies of human oocytes may help to detect correlations between predisposition to aneuploidy and putative non-disjunction inducing factors. For instance, it may be informative to examine oocytes with respect to the hormonal pretreatment of the female, and culture condition during IVF. Apart from the influence of maternal age, the possible genetic background predisposing individuals to certain structural alterations could be explored.

So far, our studies using the CBA mouse as a model, have shown that chromosomal displacements in metaphase II may be correlated to a preceeding or following non-disjunction event. Correlations between the rate of chromosome displacecment and predisposition to aneuploidy have already been postulated (Ford and Lester, 1982). However, the spreading procedures which are usually employed to assess chromosome displacement have the tendency to disperse chromosomes instead of preserving their _in vivo_ alignment. This is particularly evident in mammalian oocytes spread by the Tarkowsky method (Sugawara and Mikamo, 1986). Besides, alterations in short lived mitotic cells may not be typical for long living, dictate-arrested oocytes. In contrast, the cytoskeleton and chromosome order of the oocyte itself are well preserved by immunofluorescence. Therefore such examination may be more useful to reveal predisposition to aneuploidy by showing disturbancies in chromosome alignment.

Our data on chromosome displacement in metaphase II oocytes corroborate the notion that it is the physiological age of the ovary and its depletion rather than the age of the female _per se_ which could predispose to aneuploidy (Brook et al. , 1984). Furthermore, studies on the mouse model imply that differences in cell cycle regulation during maturation of the oocyte are cruical for aetiology of non-disjunction. Either shortening of prometaphase leading to wrong orientation of bivalents, or premature anaphase trigger causing bivalents still located in one spindle half to remain unseparated, could be responsible for loss of controlled segregation of homologes to opposite spindle poles.

Since the differences in hyperploidy and in oocyte maturation are expressed after isolation of the cells from their natural environment and _in vitro_ culture under identical conditions, they must reside in intrinsic properties of the oocyte rather than in any disturbancies caused by alterations in the immediate environment of the oocyte during its final stages of maturation as has previously been postulated(e.g. Kaufmann, 1985). Hypotheses relating environmental damage to age-related aneuploidy are less probable in view of the importance of the physiological age of the ovary opposed to the chronological age of the female, and the mean period of time an oocyte rests in the ovary before resumption of meiosis. Such observations certainly do not exclude the existence of a "production line" (Henderson and Edwards, 1968) in wich the sequential activation of oocytes and their entry into the final stages of meiosis could be predetermined. However, a cell with reduced chiasmata and a greater number of univalents, as predicted to be present in oocytes of aged females by the "production line" hypothesis, is expected to react much more sensitively, or at least equally responsive, to a further disturbance by a microtubule-disrupting drug. In contrast, the rates of aneuploidy are similar after nocodazole treatment of oocytes in all age-groups, and relative increases of hyperploidy in oocytes are rather low in old compared to young mice. Evidence presented here strongly suggests that it is the altered timing and progression through meiosis which is responsible for the increase in non-disjunction in the aged female, and not the elevated levels of univalents.

Which events and mechanisms could, in turn, be responsible for alterations in the triggering of GVBD, spindle formation, anaphase and cytokinesis in oocytes of aged individuals? From recent studies we know that the activation of MPF (e.g. Newport and Kirschner, 1984; Lohka and Maller, 1985), protein phosphorylation/dephosphorylations (e.g. Schultz et al., 1983), and ionic fluxes (reviewed in Wolniak, 1988) are intimately involved in control of the cell cycle. Also, the condition of the oocyte is greatly influenced by hormone-controlled feedback mechanisms acting during and on folliculogenesis. Possibly, such feedback mechanisms become altered due to changes in cell population in the nearly depleted ovary (reviewed in Finch and Gosden, 1986), which causes the oocyte to develop faster after release from dictate (Polanski, 1986). We have started to analyze protein synthesis, -phosphorylation, and gene activity in murine oocytes with regard to ageing and hormonal pretreatment. We hope that such studies together with immunofluorescence studies and cytogenetic screening will provide some insight into the molecular mechanisms which are responsible for altered behaviour, and ultimately play a pivotale role in predisposition to non-disjunction.

ACKNOWLEDGEMENTS:

I am most grateful to Dr. A.C. Chandley, Dr. R.G. Gosden, Dr. A. Stahl, Dr. J.M. Luciani and Dr. R. Eichenlaub without whose expertise, interest, encouragement and financial as well as moral support the current study could not have been performed. Antibodies were kindly provided by Dr. J. Kilmartin, Dr. B. Jockusch, and Dr. B. Jeppesson. The expert technical assistance of B. Tresser and typing of the manuscript by B. Obasohan is acknowledged. The work on nocodazole treatment has been performed by I. Boll as part of the requirements for a Diploma thesis. The work has been supported by a grant from the DFG (SFB 223).

REFERENCES

Balczon R, Schatten G (1983). Microtubule-containing, detergent-extracted cytoskeletons in sea urchin eggs from fertilization through cell division: anti-tubulin immunofluorescent microscopy. Cell Motil 3: 213-226.

Boll I, Eichenlaub-Ritter U (1988).Mechanisms of induction of aneuploidy with regard to ageing: sensitivity of mammalian oocytes towards nocodazole. Eur J Cell Biol 46 (Suppl 22): 10 (Abstr).

Bond DJ, Chandley AC (1983). Aneuploidy. Oxford University Press, Oxford.

Brook JD, Gosden RG, Chandley AC (1983). Maternal ageing and aneuploid embryos - evidence from the mouse that biological and not chronological age is the important influence. Hum Genet 66: 41-45.

Eichenlaub-Ritter U, Chandley AC, Gosden RG (1986). Alterations to the microtubular cytoskeleton and increased disorder of chromosome alignment in spontaneously ovulated mouse oocytes aged _in vivo_: an immunofluorescent study. Chromosoma 94: 337-345.

Eichenlaub-Ritter U, Chandley AC, Gosden RG (1988). The CBA mouse as a model for age-related aneuploidy in man: studies of oocyte maturation, spindle formation and chromosome alignment during meiosis. Chromosoma 96: 220-226.

Eichenlaub-Ritter U, Stahl A, Luciani JM (1988). The microtubular cytoskeleton and chromosomes of unfertilized human oocytes aged _in vitro_. Human Genet, 80: 259-264.

Fabricant JD, Schneider E (1978). Studies on the genetic and immunologic components of the maternal age effect. Dev Biol 66: 41-45.

Finch CE, Gosden RG (1986). Animal models for the human menophause. In: Ageing, reproduction, and the climacteric. Eds. L Mastroianni, CA Paulsen, Plenum Publ Corp, pp 3-34.

Ford JH, Lester P (1982). Factors affecting the displacement of human chromosomes from the metaphase plate. Cytogenet Cell Genet 33: 327-332.

Gianfortoni JG, Gulyas BJ (1985). The effects of short term incubation (ageing) of mouse oocytes on _in vitro_ fertilization, zona solubility and embryonic development. Gamete Res 11: 59-68.

Gosden RG (1973). Chromosome anomalies of pre-implantation mouse embryos in relation to maternal age. J Reprod Fertil 35: 351-354.

Hassold TJ, Jacobs PA (1984). Trisomy in man. Annu Rev Genet 18: 69-97.

Hassold TD, Chiu D, Yamane J (1984). Parental origin of autosomal trisomies. Ann Hum Genet 48: 129-144.
Henderson SA, Edwards RG (1968). Chiasma frequencies and maternal age in mammals. Nature 218: 22-28.
Hoebeke J, van Nijen G, De Brabander M (1976). Interaction of nococazole (R 17934), a new antitumoral drug, with rat brain tubulin. Biochem. Biophys. Commun. 69: 319-324.
Jacobs PA, Hassold TJ, Whittington E, Butler G, Gollyer S, Keston M, Lee M (1988). Klinefelder's syndrome: an analysis of the origin of the additional sex chromosome using molecular probes. Ann Hum Genet 52: 93-109.
Kaufmann MH (1985). An hypothesis regarding the origin of aneuploidy in man: indirect evidence from an experimental model. J Med Genet 22: 171-178.
Kilmartin JV, Wright B, Milstein C (1982). Rat monoclonal antitubulin antibodies derived using a new secreting rat cell line. J Cell Biol 93: 576-582.
Lohka MJ, Maller JL (1985). Induction of nuclear envelope breakdown, chromosome condensation, and spindle formation in cell-free extracts. J Bell Biol 101: 518-523.
Maro B, Howlett KS, Webb M (1985). Non-spindle microtubule-organizing centers in metaphase II-arrested mouse oocytes. J Cell Biol 101: 1665-1672.
Martin RH, Dill FJ, Miller JR (1976). Nondisjunction in ageing female mice. Cytogenet Cell Genet 17: 150-160.
Newport JW, Kirschner MW (1984). Regulation of the cell cycle during early Xenophus development. Cell 37: 731-742.
Pelestor F, Sele B (1988). Assessment of aneuploidy in the human female using cytogenetics of IVF failures. Am J Genet 42: 274-283.
Placot M, Veiga A, Montagut J, DeGrouchy J, Calderon G, Lepretre S, Junca AM, Sanatalo J, Charles E, Mandelbaum J, Barri P, Degoy J, Cohen J, Egozcue J, Sabatier JC, Salat-Baroux J (1988). Are clinical and biological IVF parameters correlated with chromosomal disorders in early life: a multicentric study. Hum Reprod 3: 627-635.
Polanski Z (1986). _In vitro_ and _in vivo_ maturation rate of oocytes from two strains of mice. J Reprod Fert 78: 103-109.
Schatten, G, Simerly C, Schatten H (1985). Microtubule configurations during fertilization, mitosis, and early development in the mouse and the requirements for egg microtubule-mediated motility during mammalian fertilization. Proc Natl Acad Sci USA 82: 4152-4156.

Schultz RM, Montgomery R, Belanoff JR (1983). Regulation of mouse oocyte meiotic maturation: implications of a decrease in oocyte cAMP and protein dephosphorylation in commitment to resume meiosis. Dev Biol 97: 264-273.

Sugarara S, Mikamo K (1986). Maternal ageing and non-disjunction: A comparative study of two chromosomal techniques on the formation of univalents in first meiotic metaphase oocytes of the mouse. Chromosoma 93: 321-325.

Szöllösi D, Mandelbaum J, Placot M, Salat-Baroux J, Cohen J (1986). Ultrastructure of the human preovulatory oocyte. J in vitro Fertil Embryo Transfer 3: 232-242.

Szöllösi D (1975). Mammalian eggs ageing in the fallopian tubes. In: Blandeau RJ (ed.). Ageing gametes. Karger, Basel pp 98-121.

Tease C, Fisher G (1986). Oocytes from young and old mice respond differently to colchicine. Mutat Res 173: 31-34.

Wassarman PM, Fujiwara K (1978). Immunofluorescent anti-tubulin staining of spindles during meiotic maturation of mouse oocytes in vitro. J Cell Sci 29: 171-188.

Wolniak SM (1988). The regulation of mitotic spindle function. Biochem Cell Biol 66: 490-514.

Wramsby H, Fredga K (1987). Chromosome analysis of human oocytes failing to cleave after insemination in vitro. Hum Reprod 2: 137-142.

Molecular and Cytogenetic Studies of
Non-Disjunction, pages 221–244

NON DISJUNCTION AND MATERNAL AGE IN THE MOUSE

Anna Estop

Department of Medical Genetics
West Penn Hospital
Pittsburgh, PA 15224

INTRODUCTION

Increasing maternal age in humans, as well as experimental animals leads to a decline in reproductive capacity. This decline is accompanied by an increased frequency of fetal death and an increased number of offspring with chromosome anomalies. Numerous hypotheses have been proposed to account for the association between maternal age and the risk of conceiving a trisomic fetus. Years ago, Polani et al. (1960), proposed the failure of dissolution of the nucleolus in aged oocytes as the possible mechanism leading to non disjunction in the D- and G- groups of human chromosomes. The "production line" is another hypothesis (Henderson and Edwards, 1968) based on an observation of a lower chiasma frequency and an increase in the number of univalents in Metaphase I (MI) oocytes of older female mice. However, Polani and Jagiello (1976) found that in older female mice, there was no parallelism between the frequencies of univalents at MI and the errors identified at MII. Furthermore, Sugawara and Mikamo (1986) concluded that univalents can also be formed as artifacts after comparing frequencies of univalents in oocytes prepared by two different methods. Delayed ovulation might affect the incidence of spindle defects in overripe eggs (Sugawara and Mikamo, 1980) and also the asynchrony between the terminalization of chiasma and chromosome alignment on the spindle (Crowley et al., 1979). Finally, the irregular cyclicity and hormonal imbalance characteristic of the premenopausal period could be a cause of chromosome anomalies (Lyon and Hawker, 1973). The hypothesis of Hansmann (1983) states that "alterations at the level of the gonadal-pituitary-hypothalamic axis for the endocrine control of follicular function may increase the probability of meiotic non disjunction," relating non disjunction to the physiology of the reproductive system. It has been known for a while that seventy percent of trisomic 21 fetuses are the result of an error in the first meiotic division (Mattei et al., 1979). However, errors in the second meiotic division also take place and those have been related to delayed fertilization: i.e. Juberg (1983) found evidence of delayed fertilization in mothers of trisomy 21 babies. An association between delayed fertilization and an important increase in chromosome anomalies has been reported by Plachot et al. (1988) in human in vitro fertilized embryos.

The laboratory mouse has been used as an experimental model to study the effects of maternal ageing. However, the maternal age effect in this model has been a matter of controversy (Golbus, 1981; Beerman et al., 1987). Different strains, females of different ages, different oocyte maturation methods and different stages of mouse development have been employed leading to an array of confusing results.

In this paper, data will be reviewed on the incidence of aneuploidy in ovulated mouse oocytes, in fertilized oocytes, and in preimplantation and postimplantation embryos. Evidence is presented for a maternal age effect for at least some strains of mice.

ANALYSIS OF ANEUPLOIDY IN METAPHASE II OOCYTES

The occurrence of hyperhaploidy in ovulated oocytes (MII stage of meiosis) is a direct measure of errors in meiosis I. The rates of hyperhaploidy reported in different papers have been chosen as a measure of aneuploidy. This is due to the fact that most authors observe higher rates of hypohaploidy than hpyerhaploidy, both in ovulated oocytes as well as in embryos (Röhrborn, 1972; Martin et al., 1976; Brook et al., 1984; Santalo et al., 1986; etc.), while other authors do not even report hypohaploidy (Fabricant and Schneider, 1978; Catala et al., in press). The preponderance of hypohaploidy is probably due to chromosome loss. A conservative estimate of the frequency of non disjunction can be obtained by doubling the hyperhaploidy (oocyte) or hyperploidy (embryo) rate.

The lack of influence of gonadotropins (GT) on the incidence of female mouse non disjunction was established by Röhrborn (1972) and Hansmann and El-Nahass (1979). The former found similar levels of hyperhaploidy in hormonally induced (2.4%) and in spontaneously obtained (2.3%) C3H oocytes. In the same way, Hansmann and El-Nahass (1979) found similar levels of hyperhaploidy, between 0.2 and 1%, in oocytes obtained by spontaneous ovulation and in gonadotropin induced cycles in 8 to 12 week old females of different strains of mice (see table 1). Superovulation did not increase aneuploidy either in young CBA (0.7% hyperhaploidy in controls and 0.7% in gonadotropin induced females) nor in older females (1% in spontaneously ovulated and 2% in stimulated females). The same pattern was observed for Swiss-Webster females in the same age groups (Golbus, 1981). Gonadotropin induced ovulation was also studied by de Boer and Van der Hoeven (1980). They reported 0.3% of hyperhaploidy in 13 to 17 week old and 0.8% in 48 to 60 week old female mice homozygous for the T(1;13)70H reciprocal translocation. A slight maternal age effect was observed, though not enough to withstand statistical evaluation.

When oocytes were cultured in vitro and analyzed at MII stage, Polani and Jagiello (1976) reported 1.2% hyperhaploidy in prepubertal (2 to 4 1/2 weeks old) CS1 females as compared to 0% in adult and aged females. In the CFLP strain they found 0.4% hyperhaploidy in older females in contract to 0% in prepubertal and young adults. The number of analyzed oocytes in each age group was relatively small in this study and the efficiency of the maturation method employed was not reported. Martin et al. (1976) also used in vitro maturation in order to study MII oocytes from CBA females. They observed 5.2% hyperhaploidy rate in "middle aged" mice (21 to 34 weeks) and none in the younger (8 to 21 weeks) or older (34 to 47 weeks) females. These data might be difficult to interpret for this high rate of hyperhaploidy in the middle age group. The special reproductive characteristics of the CBA strain must be taken into account since their fertility decreases significantly around 6 months of age and is completely impaired at 10 to 11 months (Jones and Krohn, 1961). In terms of reproductive potential, therefore, those CBA females were similar to older females of other strains used by other investigators. Martin et al., (1976) postulate that the oocytes of the oldest females may have many univalents. Consequently, they fail to progress to metaphase II and therefore become atretic. Golbus (1981) studied MII oocytes matured by different methods in Swiss-Webster random bred and in CBA mice. The results from the two strains were remarkably similar despite the differences in reproductive life span. In vitro maturation resulted in increased hyperhaploidy rates for oocytes from younger females (6.2% in in vitro matured oocytes vs. 0.7% in spontaneously ovulated) and also in older females (3% in vitro vs. 1% in spontaneous ovulation) in CBA and (2.5% and 1.2% respectively in young Swiss-Webster). This marked effect of in vitro maturation in oocytes of CBA females in this report contrasts with the data of Martin et al. (1976) where hyperhaploid in vitro matured oocytes were found only in the "middle aged" group of females. Regarding a maternal age effect in the data presented by Golbus (1981), a trend in increasing numbers is observed. For example, in spontaneously ovulated CBA's, hyperhaploidy ranges from 0.7% in 6 to 8 week old females to 1.0% in 40 to 44 week old. Also in superovulated females, hyperhaploidy increases from 0.7% to 2.0% in the two age groups (statistically significant $p<0.02$). A similar pattern is also found in Swiss-Webster: In spontaneous ovulation, oocyte hyperhaploidy ranges from 1.2% in younger females to 1.8% in older ones. In superovulated females, hyperhaploidy varies from 1.8% to 3.1% in the two groups of age. Beerman et al. (1987) described contrastingly different results in T(2;4) 1Gö/+ mice. Hyperhaploidy was highest in 18 to 22 week old females, as compared to older females which did not show any hyperhaploid oocytes. The sample size of this study (57 cells in older females) was rather low.

Another cause of hyperhaploidy other than non disjunction is presegregation. The occurrence of anaphase I presegregated chromatids can be observed only in MII oocytes. However, it is a phenomenon noted only by some authors (Hansmann and El Nahass, 1979; deBoer and Van der Hoeven, 1980; Beerman et al., 1987) and not by others who also did chromosome analysis in MII oocytes (Martin et al., 1976; Golbus, 1981; Hansmann and Jenderny, 1983). Presegregated chromatids are caused by the equational separation of a chromosome during anaphase I and result in the ovulation of oocytes with single, unpartnered chromatids. Presegragation produces hyperhaploid and hypohaploid germ cells. There is some evidence from two different authors that small chromosomes might be more prone to presegregation.

TABLE 1 Incidence of hyperhaploidy and presegregation in MII oocytes from different strains and age groups

Reference	Strain	Age (Wks)	n° cells analyzed	Hyper-haploid	% Hyper-haploid	Pre-segrega-tion	maturation method
Röhrborn, G 1972	$(101xC3H)F_1$	-	81	-	6.2	-	GT
	C3H	-	335	-	2.4	-	GT
		-	125	-	2.3	-	-
Hansmann and El-Nahass 1979	C3H/HeHan	8-12	413	1	0.2	2	-
		8-12	440	1	0.2	1	GT (low)
		8-12	451	2	0.4	2	GT (high)
	$(101xC3H)F_1$	8-12	329	1	0.3	2	-
		8-12	226	1	0.4	1	GT (low)
		8-12	187	1	1.0	2	GT (high)
	NMRI/Han	8-12	438	1	0.2	1	-
		8-12	509	1	0.2	3	GT (low)
		8-12	423	1	0.2	1	GT (high)
Golbus, 1981	CBA	6-8	275	17	6.2	-	in vitro
		40-44	200	6	3.0	-	"
		6-8	275	2	0.7	-	-
		40-44	200	2	1.0	-	-
		6-8	275	2	0.7	-	GT
		40-44	200	4	2.0[a]	-	"
	Swiss-Webster	6-8	450	11	2.5	-	in vitro
		52-64	225	8	3.5	-	"
		6-8	350	4	1.2	-	-
		52-64	175	3	1.8	-	-
		6-8	350	4	1.2	-	GT
		52-64	225	7	3.1	-	"

(Continued)

TABLE 1 (cont'd.)

Reference	Strain	Age (Wks)	n° cells analyzed	Hyper-haploid	% Hyper-haploid	Pre-segrega-tion	maturation method
Martin et al, 1976	CBA	8-21	187	0	0	-	in vitro
		21-34	116	6	5.2[b]	-	"
		34-47	155	0	0	-	"
Polani and Jagiello, 1976	CSI	2-4.5	84	1	1.2	-	in vitro
		9-27	170	0	0	-	"
		36-74	133	0	0	-	"
	CFLP	4-4.5	34	0	0	-	"
		9-27	94	0	0	-	"
		36-74	243	1	0.4	-	"
de Boer and Van der Hoeven, 1980	homT(1;13)70H	13-17	613	2	0.3	1	GT
		48-60	490	4	0.8	5	"
Beerman et al, 1987	het.T(2;4)1Gö/+	18-22	114	-	1.8	4.4%	GT low
		18-22	136	-	0.7	0.7%	GT high
		40-56	57	-	0	1.8%	GT low

a - Statistically significant p<0.02

b - Statistically significant when compared to old maternal age group p<0.025, and when compared to young maternal age group p<0.01

Polani and Jagiello (1976), observed two cells with two extra small chromatids. Beerman et al., (1987) noted that in heterozygous female mice for the translocation T(2;4) the small marker chromosome 4^2 was preferentially involved in presegregation. Chromosome analysis of one-cell mouse embryos and either preimplantation or postimplantation stage does not allow one to differentiate between hyperhaploidy resulting from meiotic non disjunction and hyperhaploidy from meiotic presegregation. Data on the relationship between maternal age and presegregation are inconclusive.

HYPERHAPLOIDY ANALYSIS IN ZYGOTES

An analysis of the chromosome complements of one-cell mouse zygotes provides a good method for determining the effect of maternal age on the primary incidence of aneuploidy. At this stage, the egg-derived chromosomes can be distinguished from the sperm complement because of their differential condensation (Donahue, 1972). Subsequent to sperm penetration and complementation of meiosis in the egg, the male and female nuclei form pronuclei, replicate their DNA, and then enter first cleavage prophase. The two prophase chromosome groups remain physically separate until they continue to form first cleavage metaphase. If an antimitotic is applied after fertilization, the two chromosome sets will not reach syngamy and they will remain in separate clusters with the female derived chromosomes being more condensed than those of the male (figure 1).

Meiotic errors (both MI and MII) will accumulate in the egg and will be apparent at the first cleavage stage. At this level, the primary incidence of aneuploidy has not suffered embryonic selection yet. An additional advantage is that every embryo produces one single spread composed of two clusters, whereas analysis in latter stages of development becomes more complex due to the presence of a higher number of spreads per embryo, which many times leads to overlapping between chromosome spreads and a false incidence of mosaics.

This system has been used by different authors to determine the primary incidence of aneuploidy in female mice. Gonadotropin stimulation was used in all reports. The results are summarized in table 2. Maudlin and Fraser (1978) found a significantly increased maternal incidence

FIG. 1 1-cell embryo at metaphase of the 1st cleavage division. Maternal and paternal pronuclei are identified.

TABLE 2 Female hyperhaploidy in one-cell mouse zygotes from different age groups.

Reference	Strain	Age (wks)	n° cells analyzed	Hyper-haploid	% Hyper-haploid
Maudlin and Fraser, 1978	TO	9-18	434	4	0.9
		35-43	375	12	3.2[a]
Tease, 1982	(CH/HeHx 101/H)F1	10-14	522	1	0.2
		43-54	324	5	1.5[b]
Martin de Leon and Boice, 1983	Swiss ICR	9-15	314	5	1.6
Catala et al, in press	(C57Blx CBA/Ca)F_1	3,5-4	537	15	2.8[c]
		10-12	590	4	0.7
		24-28	567	12	2.1[d]

[a] - Statistically significant p<0.05
[b] - Statistically significant p=0.033 (Fisher exact text)
[c] - Statistically significant p<0.005 from young adult females
[d] - Statistically significant p<0.02 from young adult females

of hyperhaploidy in aged TO random bred females: 3.2% vs. 0.9% in young adult female mice. The old females in this study had reached the age when reproductive efficiency had begun to decline in this strain (8 to 10 mo.). The results of Martin de Leon and Boice (1983) in young adult outbred ICR female mice are in the same range. Their hyperhaploidy level was 1.6%. Tease (1982) described female hyperhaploidy in F_1 hybrids from C3H/HeH x 101/H, also using first cleavage zygotes. He found statistically significant differences in hyperhaploidy in 10 to 14 week old females (0.2%) and in 43 to 54 week old females (1.5%). Data on hyperhaploidy in one-cell mouse embryos seems to be consistent among different authors after the appropriate corrections are made considering the critical age in which a decline in fertility is observed in the different strains employed by different investigators.

The influence of maternal age, including immature females, on the incidence of aneuploidy and hence hyperhaploidy was studied in first cleavage mouse embryos from (C57Bl x CBACa) F_1 hybrids by Estop et al. (1986) and Catala et al. (in press). The age of the females ranged from 3.5 to 4 weeks (immature or prepubertal) to 10 to 12 weeks (young adult) to 24 to 28 weeks (aged) females. Female mice of more than 6 months were considered aged because the average litter size declines steadily after this age. Our results suggest a high incidence of hyperhaploid oocytes in immature (2.8%) and aged female mice (2.1%) in contrast to young adult 0.7%. Statistical differences occur when comparing the incidence of hyperhaploidy in the lower and young adult age groups ($p<0.005$) and between young adult and older females ($p<0.02$). No statistical differences between the number of hyperhaploid oocytes in the first and third groups were observed. The frequency of hyperploid embryos follows the same pattern as the frequency of hyperhaploid oocytes in the three groups of age, demonstrating that variations in embryo hyperploidy are due to the age of the female when males of the same age are used (figure 2).

In the study of Catala et al. (in press), we report in vivo fertilization rates of 74.72% in immature females, 69.53% in young adults, and 51.97% in aged females. As the fertilization rate decreases, the number of degenerated eggs increases (from 7.72% to 23.86% and 41.35% respectively). The number of fertilized eggs is calculated

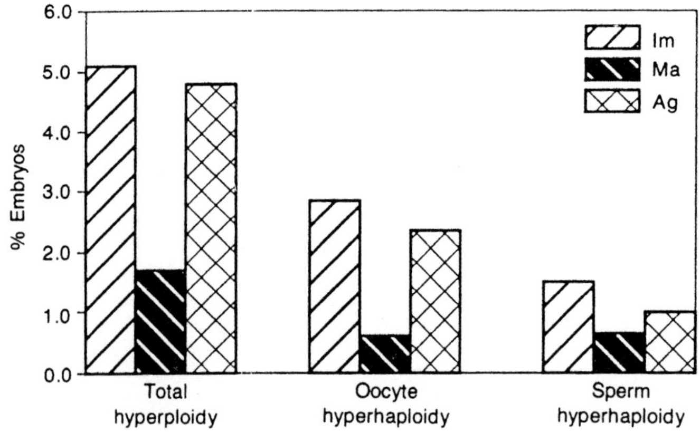

FIG. 2 Hyperploidy in 1st cleavage mouse embryos from prepubertal (Im), young (Ma) and aged (Ag) females.

at the time of chromosome analysis and at that time, all degenerated eggs are scored as such and consequently as unfertilized. Some of those, however, may have been already degenerated at the time of ovulation because they were in some way damaged, or because they had too many univalents, and then they become atretic. Hence, the incidence of hyperhaploidy in immature females should be considered as the most accurate incidence since the number of eggs that are fertilized and therefore analyzable is highest. In young adults, some atretic oocytes (with meiotic block) could account for the lowering of the fertilization rate. In older females, higher numbers of atretic oocytes (again with meiotic block) could account for a lower fertilization rate and could also imply that the hyperhaploidy in this age group is underestimated. Oocytes remaining in the oviduct from a previous ovulation cannot contribute to increase the numbers of degenerated oocytes because ovulated oocytes are picked up in the ampullae where they are all clustered together with the cumulus cells and the oviducts were not flushed. This decline in fertility would not be the result of an increased failure of implantation nor a failure in development, but the result of a lower number of healthy oocytes.

The results from first cleavage chromosome analysis show evidence of a maternal age effect in non disjunction in three strains of mice: TO, (C3H/HeH x 101/H)F_1 and

(C57Bl x CBA/Ca)F_1. A high incidence of non disjunction in gonadotropin induced immature females might be a generalized phenomenon in mammals.

ANALYSIS IN PREIMPLANTATION AND POSTIMPLANTATION EMBRYOS (table 3)

A very limited number of morulae and blastocysts (3½ days after coitus) were studied in CBA/HT6 female mice of 1 to 12 months of age by Gosden (1973). A maternal age effect was observed starting with 2.1% of trisomic embryos in the 1 month old females (1/47), 0% in 2 to 7 months to 6.9% in older females (8 to 12 months). In CBA females, an earlier cessation of reproduction, brought on by unilateral ovariectomy, resulted in the earlier onset of irregular cyclicity and an earlier rise in aneuploidy (Brook et al., 1984). Hyperploidy was not increased in the younger group after ovariectomy but in the "middle aged" (5 to 6 months) and older (8 to 9 months) a rise in trisomy (4% vs. 0%) and (5% vs 1.7%) was observed in ovariectomized females as compared to those who were sham operated. Due to the small sample size, trisomy values do not reach significance. These very interesting results suggest that the size of the follicle population limits the length of the functional ovarian lifespan in CBA mice and it is conceivable to assume that this is a general phenomenon in mammals. Acyclicity would result when a critical threshold number of large secretory follicles, required for maintaining the positive feedback mechanism for pituitary gonadotropin release, is not obtained.

In postimplantation embryos, a consistent finding is an increase in the number of trisomic embryos with higher maternal age. Yanamoto et al. (1973a) did not find any trisomic embryos (10.5 days post coitus) in young CFI mice of 13 to 22 weeks; however, in older mice (48 to 52 weeks) and in (60 to 69) weeks, they observed 1.7% and 2.6% of trisomic embryos. In the same manner, Fabricant and Schneider (1978), using many strains of different ages, observed a rise in the frequency of trisomic embryos with maternal age from 0% to 3%. The most striking increase was observed in the CBA strain, where the number of trisomic embryos rose from 0/45 (0%) in 8 to 22 week old females to 3/44 (6.8%) in 30 to 44 week old females. These chromosome studies were carried out between the tenth and fourteenth

TABLE 3 Hyperploidy in pre- and postimplantation mouse embryos.

Reference	Strain	Age (wks)	Stage	n° cells analyzed	Hyperploid	% Hyperploid	GT
Gooden, 1973	CBA/HT6	4-4.5	mo-blast	47	1	2.1	+
		18-52	mo-blast	63	-	0.0	+
		60-69	mo-blast	58	4[a]	6.9[b]	+
Brook et al, 1984	CBA/Ca	9-11 (ovx)	mo-blast	67	-	0	-
		9-11 (shm)	3.5 days	66	-	0	-
		22-26 (ovx)	"	75	3	4	-
		22-26 (shm)	"	65	-	0	-
		35-40 (ovx)	"	41	2	5	-
		35-40 (shm)	"	58	1	1.7	-
		44-50 (shm)	"	46	1	2,2	-
Yanamoto et al, 1973a	CF1	13-22	Postimplant.	149	-	0	-
		48-52	10.5 days	117	2	1.7	-
		60-69		39[c]	1	2.6	-
Fabricant and Schneider, 1978	CBA,NZBL.	8-22	Postimplant.	239	0	0	-
	C57Bl/6J	30-44	10-14 days	166	5	3	-
	A/J,$_c$C3H/HeJ	>48		111	2	1.8	-

[a] - One embryo with 42 hr

[b] - Statistically significant P=0.0495 (Fisher's exact test)

[c] - Includes 1 mosaic 40/41; 1 mosaic 39/40/41

days of gestation and a relatively low number of embryos of each strain were used.

MATERNAL AGE SENSITIVITY TO INDUCED MEIOTIC NON DISJUNCTION

It is assumed that the maternal age relationship to aneuploidy is secondary to an increased risk of non disjunction. Some human epidemiological studies (Uchida et al., 1968; Alberman et al., 1972) suggest that there may be an association between non disjunction and preconception exposure to abdominal X-rays in women. This response to radiation seems to increase with age (Alberman et al., 1972). X-rays, plus a series of suspected chromosomal mutagens have been used in the mouse model to test whether the incidence of induction of aneuploidy is a function of maternal age.

Uchida and Freeman (1977) used X-rays in an attempt to induce meiotic non disjunction in oocytes from (C3H x ICR/Swiss) F_1 hybrids. Females were irradiated an average of 20 rads one week prior to sacrifice and the cells were analyzed at MII. Irradiated females had higher levels of hyperhaploidy in comparison to control mice, both in aged (12 mo.) and in young mice (3 to 6 mo.). (table 4) When comparing hyperhaploidy in old irradiated females, (1.3%), and old controls, (0.3%), statistical significances appear ($p<0.01$), whereas hyperhaploidy in young irradiated females

TABLE 4 The effect of radiation on the induction of hyperhaploidy in the mouse

Reference	Strain	Age (wks.)	n° cells analyzed	Hyper-haploid	% hyper-haploid	Stage analyzed	Stage/ Radiation
Uchida and Freedman, 1977	(C3Hx ICR/Swiss)	13-26	1054	0	0	MII	--
		13-26	1149	6	0.5	MII	dictyate/20 rad
		52	1036	4	0.3	MII	--
		52	1117	15	1.3	MII	dictyate/20 rad
Tease, 1982	(C3H/HeHx) 101/H)	10-14	523	1	0.2	1st cleavage	--
		10-14	504	6	1.2	"	Diak-MI/10 rad
		10-14	508	10	2.0	"	Diak-MI/25 rad
		10-14	504	17	3.4	"	Diak-MI/50 rad
		43-54	324	5	1.6	"	--
		43-54	268	7	2.6	"	Diak-MI/10 rad
		43-54	287	6	2.0	"	Diak-MI/25 rad
		43-54	280	14	5.0	"	Diak-MI/50 rad

(0.5%) is not statistically significant from young female controls (0%). They concluded that radiosensitivity dependent non disjunction increases with age. These results contrast sharply with those of Tease (1982) who found similar dose-related chromosome non disjunction in young and old (C3H/HeH x 101/H)F_1 female mice after X-irradiation. Here, the analysis was done in first cleavage mouse embryos obtained from females exposed to similar doses as the previous reference. However, the stage of exposure was 3 hours post HCG (Diakinesis/MI). Significant differences occur between young and old mice in the levels of non disjunction due to a greater spontaneous incidence in old mice. The increase in hyperhaploidy due to radiation-induced non disjunction was identical in young and old females indicating no differences in sensitivities of their oocytes to X-rays.

Yanamoto et al. (1973b) found an age dependent increase in sensitivity to radiation induced aneuploidy; however, most of the aneuploids described were actually mosaics which most probably originate from mitotic non disjunction. Attempts to repeat the experiments have failed to confirm the findings (Strausmanis et al., 1978) and a revision of the statistics reported by Yanamoto on the interaction between age and X-irradiation, proved that there was no evidence of an interrelation between maternal age and radiation treatment (Gosden and Walters, 1974).

Variables such as mouse strain and oocyte maturation stage at the time of irradiation may account for the equivocal conclusions drawn by different authors. Oocytes at different stages of development and maturation have different response characteristics to irradiation (Brewen et al., 1976, Brewen and Payne, 1979). Hansmann (1974), for example, has looked at stage sensitivity of mouse oogenesis to amethopterin and cyclophosphamide and found that induction of aneuploidy is related to highly sensitive stages when chromosome segregation takes place, such as in oogonia and during the immediately preovulatory (Diakinesis/MI) stage. Dictyate treated with X-rays 1 week prior to superovulation is relatively insensitive (Tease, 1981). The age relation to non disjunction does not seem to be related to a greater sensitivity of oocytes from older females to radiation. X-rays are known to cause chromosome fragments and chromatid interchanges in both oocytes and spermatocytes (Brewen et al., 1976; Chandley and Speed, 1979; Tease et al., 1982) and this effect might be much more critical than its incidence in non disjunction.

PRODUCTION OF DIPLOID OOCYTES

From data on table 5, it can be inferred that there is an absence of a maternal age effect on the production of diploid oocytes, with the exception of immature females which will be discussed later. Martin et al. (1976), Polani and Jagiello (1976), Beerman et al. (1987), Catala et al. (in press) found similar levels of diploid oocytes, even though they used females of different strains, various age groups, and various oocyte maturation methods; [i.e. in vitro maturation of preovulatory oocytes (Martin et al., 1976; Polani and Jagiello, 1976), superovulated oocytes (Beerman et al., 1987) and fertilized superovulated oocytes (Catala et al., in press).]

In contrast to aneuploidy, gonadotropin stimulation seems to increase the incidence of polyploidy. Takagi and Sasaki (1976) observed an increased incidence of digynic triploidy (19.2% vs. 3.7% in controls) in mouse embryos from the A/He strain, and the major cause of digyny was a defect in the second meiotic division of the oocyte. The PMSG (pregnant mare serum gonadotropin) and HCG (human chorionic gonadotropin) dosage used was 10 iu. Hansmann and El-Nahass (1979) reported a high incidence of diploid oocytes due to the retention of the first polar body in hormonally stimulated (10/10iu) NMRI/Han mice, (10.1% vs. 2.7% in controls), but in C3H/HeHan and F_1 (101 x C3H) the same amount of gonadotropins did not increase the production of diploid oocytes. Strains as NMRI/Han mice are catalogued as "sensitive." Beerman et al, (1987) also report a greater incidence of diploid oocytes in T(2;4)1Gö/+ mice exposed to 10/10 iu of gonadotropins (3.5%) as compared to females exposed to 1.5/1.0iu (0.7%). All oocytes were arrested at MI. It seems that 10/10iu of gonadotropins is a critical concentration that might induce the ovulation of diploid oocytes in certain strains of mice. The underlying biological phenomenon seems to be rather complex. Some strains of mice retain diploid oocytes in MI, and in other strains the preovulatory phase is not affected for they produce MII diploid oocytes. In a study with F_1 hybrids from C57Bl x CBaCa, we found very low

TABLE 5 Production of diploid oocytes in different strains of mice.

Reference	Strain	age (wks)	n° cells analyzed	stage	Diploid	% Diploid	GT
Takagi and	A/He	6-8	421	Postimplantation	81	19.2[d]	10/10
Sasaki, 1976		6-8	1406	(6.5-7.5) d	52	3.7	"
Martin et al,	CBA	9-21	353	MII	4	1.1	in vitro
1976[b]		21-34	280	"	5	2.1	"
		34-47	244	"	5	2.0	"
Polani and	CFPL	4	34	MII	0	0	in vitro
Jagiello,		9-27	94	"	0	0	"
1976		36-74	243	"	4	1.6	"
	CSI	2-4	84	"	0	0	"
		9-27	170	"	0	0	"
		36-74	133	"	1	0.7	"
Hansmann and	C3H/HeHan	8-12	451	MII	2	0.4	10/10
El-Nahass	$(101xC3H)F_1$	8-12	187	"	1	0.5	"
1979	NMRI/Han	8-12	423	"	43[a]	10.1[d]	"
		8-12	438		12[a]	2.7	-
Beerman et al,	T(2;4)	18-22	575	MII	-	0.7	1.5/1.0
1987[c]	1Gö/+	18-22	599	"	-	3.5	10/10
		40-56	196	"		1.0	1.5/1.0
Catala et al,	(C57Blx	3.5-4	537	1st cleavage	14	2.6[d]	5/5
in press	$CBACa)F_1$	10-12	590	"	2	0.3	"
		24-28	567	"	3	0.5	"

a - Most of them showed typical MI pairing configuration without polar body.
b - n° analyzed oocytes containing embryos with chromosome count accuracy of ±2.
c - All diploid oocytes arrested at MI.
d - Statistically diffrent than other groups from same author.

levels of diploid oocytes in first cleavage embryos both in young adult (0.3%) and in older females(0.5%). These values are in the same range as those described by Hansmann and El-Nahass (1979) for "non sensitive" strains of mice such as C3H/HeHan and (101 x C3H)F_1, 0.4% and 0.5%, respectively. Immature females, on the contrary, have a higher incidence of diploid oocytes (2.6%) which is statistically significant from the other age groups. Females were superovulated with 5/5 iu of gonadotropins in this experiment. Since the lack of a second polar body was apparent in some diploid one-cell embryos of digynic origin, it was concluded that defects in the second meiotic division contributed to the induction of diploid oocytes along with errors in the first meiotic division. Immature females might behave as "sensitive" to gonadotropin stimulation, as adult females of other strains do. Hansmann (1983) suggests that gonadotropins might alter the endocrine control of follicular maturation in certain strains of mice. This alteration could disrupt the development of the oocyte causing an increase in heteroploidy. XO mice products of a cross between females of C3H x 101 and males of NMRI/Han "gonadotropin sensitive" strain, ovulate a high level of diploid oocytes (24.6%) as compared to their XX littermates (1.0%) (Beerman et al., 1985). Adult XO mice stop reproductive activity earlier than their XX sibs because of a reduced oocyte pool (Lyon and Hawker, 1973).

CONCLUSIONS AND OVERVIEW

Data presented in this paper allow us to infer the relationship between maternal age and non disjunction in some strains of mice. Available data from most strains are for young adult to middle aged. Data on very young females (prepubertal) or very old (at the very end of their reproductive life) are very scarce. CBA mice show a maternal age effect with a maximum hyperhaploidy level in "middle aged" females, when reproductive fitness is getting impaired (Martin et al., 1976). TO outbred females and (C3H/HeH x101/H)F_1 hybrids also show an increased incidence of non disjunction with age (Maudlin and Fraser, 1978). Hyperhaploid oocytes in MNRI/Han mice were only detected at 50 weeks of age and not in younger or older age. (Hansmann et al., 1985) also showing maternal age effect. C57Bl x CBa/Ca F_1 hybrid immature females ovulate as many hyperhaploid oocytes as aged females in contrast to young adults. These data are suggestive of a two hump shaped slope with the highest points at a very young age and middle age groups (Catala, 1987). The possibility that such a curve might be a general phenomenon in mammals cannot be ruled out for, in Djungarian hamsters, the maternal age effect is also more pronounced in middle age when the number of malsegregated bivalents per oocyte is considered (Hummler et al., 1987).

Human data in support for this two hump-shaped slope have been described. Prenatal diagnosis after amniocentesis suggests a leveling-off, or even a decrease in the slope of at least certain trisomies at very advanced maternal ages (Ferguson-Smith, 1983). Furthermore, data from spontaneous abortions relating maternal age and chromosome size also suggests a leveling-off and even a drop-off for certain trisomies with highest maternal ages (Risch et al., 1986). A relatively high level of non disjunction in younger women was suggested by Erickson (1978) and Read (1982). A moderate increase in total trisomy frequency was observed in women under 15 years of age, primarily due to trisomy 4 (Hassold and Chiu, 1985).

Although it is not the goal of this paper to discuss in depth the mechanisms of non disjunction, it is increasingly necessary to search for a hypothesis that would be in agreement with such two-humped shape slope. The concept that age-related changes in hormone levels may be responsible for age-dependent non disjunction is not new. Crowley (1979) suggested that there is an interaction between hormonally governed rate of meiosis and the timing of chiasma terminalization. As hormone levels and the length of the cycle change with advancing age of the mother, meiosis slows down and chiasma frequencies decline. The association of advanced acyclicity with non disjunction has evidence in the mouse model (Brook et al., 1984), and although speculative, it could be hypothesized that women with irregular cycles might have more tendency to non disjunction.

The signal for induction of oocyte maturation in mammals is the surge of LH (luteinizing hormone). LH acts to promote oocyte maturation by raising follicular levels of cAMP (Ekholm et al., 1984). Oocyte maturation occurs when the transfer of cAMP from the granulosa cells to the oocyte is interrupted (Behrman et al., 1988). Both estradiol and FSH contribute to induce LH receptors in granulosa cells. Estradiol values in the early follicular phase and during the peak in the late follicular phase are lower in older women, FSH is higher in older women and LH peak values are constant regardless of age (Sherman and Korenman, 1975). If FSH is increased because of age, this could result in a speed up of the meiotic rate.

At the time of resuming meiosis, a cAMP dependent protein kinase dephosphorylates non specific protein(s) which are maintained phosphorylated during meiotic arrest (Dekel, 1988). These steps would trigger meiosis from the germinal vesicle break down (GVBD) to the circular bivalent stage. From this step onwards, continuation of meiosis is dependent on protein synthesis, and the actual release of the oocyte depends on plasminogen activator which, in turn, is induced by FSH (Hsueh et al., 1988). Alterations in hormone levels could impair any of these steps by altering the accumulation of a rate limiting substance and result in either anovulatory cycles (which are known to occur in older women) or in disturbances in the meiotic rate, namely the increase or decrease of the interval between the LH peak and ovulation.

Alterations of the meiosis rate could cause many disturbances including asynchronous segregation of bivalents and meiotic block in metaphase I. The later would lead to the ovulation of diploid oocytes. If these hypotheses were true, the primary causes of non disjunction would lay in the surrounding cells of the gonad, while meiotic chromosomes would play a passive yet albeit individual role depending on their physical characteristics and on the interaction, if present, with translocation multivalents.

Chromosomal properties, such as length, centromere position, amount of constitutive heterochromatin, and presence of nucleolar organizing regions influence the probability of non disjunction. Spindle properties also play a role, and alterations in the assembly or disassembly of the kinetochore microtubules could result in malsegregation, too. Hummler and Hansmann (1985) have postulated that the bivalents that are last to segregate are at a higher risk for non disjunction due to interference with microtubular function and those, in the

Djungarian hamster, are the later meta- and submetacentric chromosomes.

Recently, the use of molecular polymorphisms on chromosome 21 have been used to study the association between non disjunction and meiotic recombination. Warren et al., (1987) have found evidence that recombination is reduced on non disjoined chromosomes 21, whereas Stewart et al. (1988) have detected recombination in one out of three cases of maternal meiosis I error. Data on recombination and its association with non disjunction with the use of DNA markers, and studies on the endocrine and molecular control of meiosis will give further insight into the different causative agents of non disjunction.

ACKNOWLEDGEMENTS

The research reported in this paper and performed in the Department of Cell Biology, Autonomous University of Barcelona (Spain) was funded by grant CAICYT 1502/83.

REFERENCES

Alberman E, Polani PE, Fraser Roberts JA, Spicer, CC, Elliot M, Armstrong E (1972). Parental exposure to X-iradiation and Down's Syndrome. Ann Hum Genet Lond 36:195-208

Beerman F, Francke U, Hansmann I (1985). High susceptibility for diploidy in ovulated oocytes from XO mice. Hum Genet 72:323-326.

Beerman F, Bartels I, Francke U, Hansmann I (1987). Chromosome segregation at meiosis I in female T(2;4)1Gö/+ mice: No evidence for a decreased crossover frequency with maternal age. Chromosoma 95:1-7.

Behrman HR, Preston SL, Pellicer A, Parmer TG (1988). Oocyte maturation is regulated by modulation of the action of FSH in cumulus cells. In Halsetine FS, First NL (eds): "Meiotic Inhibition, Molecular Control of Meiosis," New York: A. R. Liss, Inc., pp 115-135.

Brewen, JG, Payne HS, Preston RJ (1976). X-ray induced chromosome aberrations in mouse dictyate oocytes. I. Time and dose relationships. Mut Res 35:111-120.

Brewen JG, Payne HS (1979). X-ray stage sensitivity of mouse oocytes and its bearing on dose-response curves. Genetics 91:149-161.

Brook JD, Gosden RG, Chandley AC (1984). Maternal aging and aneuploid embryos - Evidence from the mouse that biological and not chronological age is the important influence. Human Genet 66:41-45.

Catala V (1987). Edat materna i anomalies cromosomiques. Estudi de la primera divisio embrionaria. Ph.D. thesis. Universitat Autonoma de Barcelona.

Catala V, Estop AM, Santalo J, Egozcue J (in press). Sexual immaturity and maternal age. Incidence of aneuploidy and polyploidy in first-cleavage mouse embryos. Cytogenet Cell Genet.

Chandley AC, Speed RM (1979). Testing for nondisjunction in the mouse. Envir Health Perspec 31:123-129.

Crowley PH, Gulati DK, Hayden TL, Lopez P, Dyer R (1979). A chiasma-hormonal hypothesis relating Down's syndrome and maternal age. Nature Lond 280:417-419.

de Boer P, Van der Hoeven FA (1980). The use of translocation derived "marker bivalents" for studying the origin of meiotic instability in female mice. Cytogenet Cell Genet 26:49-58.

Dekel N (1988). Spatial relationship of follicular cells in the control of meiosis. In Haseltine FS, First NL (eds): "Meiotic Inhibition, Molecular Control of Meiosis," New York: A. R. Liss, Inc., pp 87-101.

Donahue RP (1972). Cytogenetic analysis of the first cleavage division in mouse embryos. Proc Natl Aca Sci 69(1):74-77.

Ekholm C, Hillensjo T, Magnusson C, Rosenberg S (1984). Stimulation and inhibition of rat oocyte meiosis by forskolin. Biol Reprod 30:357-360.

Erickson JD (1978). Down's Syndrome, paternal age and birth order. Ann Hum Genet 41:489-498.

Estop AM, Catala V, Santalo J, Egozcue J (1986). Aneuploidy, maternal age and sexual immaturity. Am J Hum Genet 39(3):A 112.

Fabricant JD, Schneider EL (1978). Studies on the genetic and immunological components of the maternal age effect. Dev. Biology 66:337-343.

Ferguson-Smith MA (1983). Prenatal chromosome analysis and its impact on the birth incidence of chromosome disorders. Brit Med Bull 39:355-364.

Golbus MS (1981). The influence of strain, maternal age, and method of maturation on mouse oocyte aneuploidy. Cytogenet Cell Genet 31:84-90.

Gosden RG (1973). Chromosomal anomalies of preimplantation mouse embryos. J Reprod Fert 35:351-354.

Gosden RG, Walters DE (1974). Effects of low dosage X-irradiation on chromosomal non disjunction in aged mice. Nature 248:54-55.

Hansmann I (1974). Chromosome aberrations in metaphase II oocytes, stage sensitivity in mouse oogenesis to amethopterin and cyclophosphamide. Mut Res 22:175-191.

Hansmann I (1983). Factors and mechanisms in non disjunction and X chromosome loss. In Sandberg A (ed): "Cytogenetics of the Mammalian X Chromosome, Part A: Basic Mechanisms of X Chromosome Behavior," New York: Alan R. Liss, Inc., pp 131-170.

Hansmann I, El Nahass E (1979). Incidence of nondisjunction in mouse oocytes. Cytogenet Cell Genet 24:115-121.

Hansmann I, Jenderny J (1983). The genetic basis of non disjunction: Increased incidence of hyperhaploidy in oocytes from F_1 hybrid mice. Human Genet 65:56-60.

Hansmann I, Bartels I, Beerman F, Caspari D, Franke U, Hummler E, Theuring F (1985): Mechanisms of non disjunction: Facts and Perspectives. In Dellarco VL, Voytek PE, Hollander A (eds): "Aneuploidy, Etiology and Mechanisms". Plenum New York and London pp 417-432.

Hassold T, Chiu D (1985). Maternal-age specific rates of numerical chromosome abnormalities with special reference to trisomy. Hum Genet 70:11-17.

Hayden TL, Crowley PH, Wilson E (1980). A model relating the incidence of meiotic trisomy to maternal age. J Theor Biol 86:123-126.

Henderson SA, Edwards RG (1968). Chiasma frequency and maternal age in mammals. Nature Lond 218:22-28.

Hsueh AJW, Liu XX, Cajander SB, Ny T (1988). Molecular mechanisms in the hormonal regulation of plasminogen activator activity in ovarian granulosa cells and cumulus-oocyte complexes. In Haseltine FS, First NL (eds): "Meiotic Inhibition, Molecular Control of Meiosis," New York: A. R. Liss, Inc., pp 227-257.

Hummler E, Hansmann I (1985). Preferential non disjunction of specific bivalents in oocytes from Djungarian hamsters (Phodopus sungorus) following colchicine treatment. Cytogenet Cell Genet 39:161-167.

Hummler E, Theuring F, Hansmann I (1987). Meiotic non disjunction in oocytes from aged Djungarian hamsters correlates with an alteration in meiosis rate but not in univalent formation. Hum Genet 76:357-364.

Jones EC, Krohn PL (1961). The relationships between age, numbers of oocytes and fertility in multiparous mice. Endocrinology 21:469-494.

Juberg RC (1983). Origin of chromosome abnormalities: Evidence for delayed fertilization in meiotic non disjunction. Hum Genet 64:122-127.

Lyon MF, Hawker SG (1973). Reproductive lifespan in irradiated and unirradiated XO mice. Genet Res 21:185-194.

Martin RH, Dill FJ, Miller JR (1976). Non disjunction in aging female mice. Cytogenet Cell Genet 17:150-160.

Martin de Leon PA, Boice ML (1983). Spontaneous heteroploidy in one cell mouse embryos. Cytogenet Cell Genet 35:57-63.

Mattei JF, Mattei MG, Ayme, Girand F (1979). Origin of the extra chromosome in trisomy 21. Hum Genet 46:107-110.

Maudlin I, Fraser LR (1978). Maternal age and the incidence of aneuploidy in first-cleavage mouse embryos after in vitro fertilization. J Reprod Fert 54:423-426

Plachot M, de Grouchy J, Junca AM, Mandlebaum J, Salat-Baroux J, Cohen J (1988). Chromosome analysis of human oocyte and embryos: Does delayed fertilization increase chromosome imbalance? Hum Reprod 3(1):125-127.

Polani PE, Briggs JH, Ford CE, Clarke CM, Berg JM (1960). A mongol girl with 46 chromosomes. Lancet I:721-724.

Polani PE, Jagiello GM (1976). Chiasmata, meiotic univalents and age in relation to aneuploid imbalance in mice. Cytogenet Cell Genet 16:505-529.

Read SG (1982). The distribution of Down's Syndrome. J Ment Defic Res 26:215-227.

Risch N, Stein Z, Kline J, Warburton D (1986). The relationship between maternal age and chromosome size in autosomal trisomy. Am J Hum Genet 39:68-78.

Röhrborn G (1972). Frequencies of spontaneous non disjunction in Metaphase II oocytes of mice. Humangenetik 16:123-125.

Santalo J, Estop AM, Egozcue J (1986). The chromosome complement of first-cleavage mouse embryos after in vitro fertilization. J In Vitro Fert & Embry Transf 3(2):99-105.

Sherman M, Korenman SG (1975). Hormonal characteristics of the menstrual cycle throughout reproductive life. J Clin Inv 55:699-706.

Stewart GD, Hassold TJ, Berg A, Watkins P, Tauzi R, Kurnit DM (1988). Trisomy 21 (Down Syndrome): Studying non disjunction and meiotic recombination by using cytogenetic and molecular polymorphisms that span

chromosome 21. Am J Hum Genet 42:227-236.
Strausmanis R, Henrikson IB, Holmberg M, Romback C (1978). Lack of effect on the chromosomal non disjunction in aged mice after low dose X-irradiation. Mut Res 49:269-274.
Sugawara S, Mikamo K (1980). An experimental approach to the analysis of mechanisms of meiotic non disjunction and anaphase lagging in primary oocytes. Cytogenet Cell Genetics 20:251-264.
Sugawara S, Mikamo K (1986). Maternal aging and non disjunction: a comparative study of two chromosomal techniques on the formation of univalents in first meiotic metaphase oocytes of the mouse. Chromosoma 93:321-325.
Takagi N, Sasaki M (1976). Digynic triploidy after superovulation in mice. Nature 264:278-281.
Tease C (1981). Chromosome non disjunction in female mice after X-irradiation. Mouse Newsl 65:22-24
Tease C (1982). Similar dose-related chromosome non-disjunction in young and old female mice after X-irradation. Mut Res 95:287-296.
Uchida IA, Holunga R, Lauler C (1968). Maternal radiation and chromosomal aberrations. Lancet 16 Nov:1045-1049.
Uchida IA, Viola Freeman CP (1977). Radiation-induced non disjunction in oocytes of aged mice. Nature 205:186-187.
Warren AC, Chakravarti A, Wong C, Slaugenhaut SA, Halloran SL, Watkins PC, Metaxotou C, Antonarakis SE (1987). Evidence for reduced recombination on the nondisjoined chromosomes 21 in Down Syndrome. Science 237:652-654.
Yanamoto M, Endo A, Watanabe G (1973a). Maternal age dependence of chromosome anomalies. Nature New Biol 241:141-142.
Yanamoto M, Shimida T, Endo A, Watanabe G (1973b). Effect of low-dose X-irradiation on the chromosomal non disjunction in aged mice. Nature New Biol 244:206-209.

Molecular and Cytogenetic Studies of
Non-Disjunction, pages 245–262

AN INVESTIGATION OF FACTORS INFLUENCING THE SURVIVAL OF TRISOMIC FETUSES IN THE MOUSE.

M. Vekemans

Centre for Human Genetics
McGill University
Montreal, Canada

ABSTRACT

The prevalence of trisomy at a given time of gestation depends on two factors: its incidence at conception and the probability that the trisomic embryo or fetus survives up to the time of observation. In mammals, little is known about the factors controlling either of these two events.

A mouse model is particularly useful for these types of studies, as it allows us to examine each event, under strictly controlled experimental conditions. Using this mouse model, a systematic analysis of the influence of various factors including maternal age, diet, as well as fetal and maternal genes, on the survival of trisomic fetuses, at different times of gestation, has been performed.

Our studies show that high post-implantation survival of trisomic fetuses is associated with a specific diet or a specific maternal gene, but not increasing maternal age. Future studies are aimed at identifying the specific genetic and environmental factors involved, and understanding their pathogenesis.

INTRODUCTION

Chromosomal disorders are extremely common as they affect about 30% of all recognized conceptions. Most of these are spontaneously miscarried but this natural selection is not complete because about 1% of newborns have a chromosomal condition (Hook, 1985).

The possibilities of primary prevention, i.e. avoiding the occurence of chromosomal aberrations, are at present very limited. Since the incidence of various chromosomal aberrations, including trisomy 21, increases with maternal age, one way to reduce the incidence of these handicaps could be to discourage pregnancy at advanced maternal age by advising older mothers of the increased risk. The decision not to reproduce is, however, a very difficult one and fetal diagnosis in early pregnancy with the possibility of selective abortion (secondary prevention) has proved a much more popular alternative for couples at risk.

The incidence of chromosomal disorders can be reduced by at most one third if fetal diagnosis is limited to mothers over 35 years old and those with an indication from parental karyotyping. However, most affected births occur, with a lower incidence, among the much larger number of infants born to younger mothers. Therefore, if the young couples who produce the majority of affected children are to benefit from fetal diagnosis, progress must be made in identifying groups of pregnancies at higher risk prior to karyotyping. The present research is an attempt to identify in the population individuals at risk for giving birth to an aneuploid offspring.

THE MODEL

In 1975, Stein et al. proposed a simple model depicting the distribution among human spontaneous abortions and livebirths, of conceptuses with and without an anomaly such as a trisomy (Figure 1). From this model, it can be inferred that the prevalence of trisomy, for example at birth, is dependent on only two factors: the incidence of trisomy at conception and the probability that such a conceptus survives until term. The model also implies that, by regulating the incidence of trisomy at conception, it is possible to analyze the factors controlling the survival of trisomic fetuses.

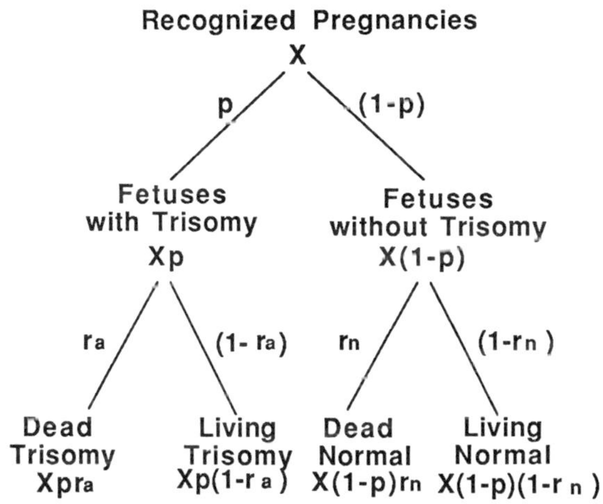

Xp = expected number of fetuses with trisomy

r_a = probability that a trisomic fetus will abort spontaneously

r_n = probability that a normal fetus will abort spontaneously

$$F = \text{Observed Rate of Trisomy} = \frac{\text{Living Trisomy}}{\text{Total Living}} = \frac{Xp(1-r_a)}{Xp(1-r_a)+X(1-p)(1-r_n)}$$

Figure 1. Distribution Among Living and Dead, of Fetuses With and Without Trisomy (Modified from Stein et al, 1975)

A trisomic state can be induced experimentally for virtually any of the 20 pairs of acrocentric mouse chromosomes (Gropp, 1982). The generation of specific types of trisomy is dependent on mouse strains carrying different Robertsonian translocations. A Robertsonian translocation involves the centric fusion of two acrocentric chromosomes. The breeding scheme used in producing trisomic conceptuses involves the mating of parental mice, each homozygous for a particular Robertsonian translocation which have one arm in common. The offspring produced from such a mating will be heterozygous for two Robertsonian translocations with partial homology for one of their arms. As a consequence, when the chromosomes partake in the first meiotic prophase, a quadrivalent is formed which often leads to the nondisjunction of the Robertsonian translocation chromosome depending on the mode of segregation during the first meiotic anaphase (Figure 2).

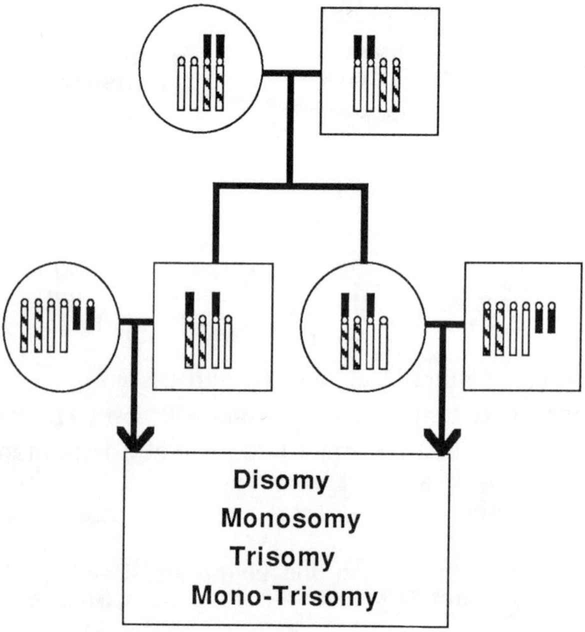

Figure 2. Breeding Design for the Production of Aneuploid Conceptuses

By following this particular breeding scheme for the production of trisomic conceptuses, the length of survival of monosomic and trisomic conceptuses was studied extensively (Gropp, 1982; Epstein, 1987). From these studies, it was clear that the fetal survival depended both on the type of aneuploidy produced and the specific chromosome involved. Also, preliminary studies suggested that it might be influenced by maternal attributes.

STUDIES OF THE SURVIVAL OF TRISOMIC FETUSES

Using the same breeding scheme for the production of aneuploid conceptuses, our laboratory has investigated in detail the possibility that the survival of trisomic fetuses might be influenced by the age, the genotype and the nutritional status of the mother.

a. Maternal Age and Survival of Trisomic Fetuses

Several hypotheses have been suggested to explain the relationship between increasing maternal age and human trisomy. One of the most provocative suggests that the maternal age effect on aneuploidy might well be due to a decreased ability to select against aneuploid embryos as a function of increasing maternal age (Ayme and Lippman-Hand 1982). Subsequent studies, however, have not supported this original proposal. For example, there was no evidence of a maternal age effect in D/21 translocation Down syndrome (Hook 1983, Carothers 1983, 1987). More recently, a mechanism of maternal screening that operates between fertilization and recognition (first zygotic division) has been postulated (Stein et al. 1986). The model suggests that in older women, the prerecognition attrition is low and that the rate of fertilization does not vary with age. Therefore, the rate of trisomy observed in miscarriages of older women is an indication of the incidence at all ages. Also, the characteristics of the individual chromosomes may play a role in the efficiency of the rejection procedure.

Hence with advanced maternal age, the small chromosomes could escape the screening process thereby contributing to a higher incidence in spontaneous abortions and livebirths. This new proposal, however, does not appear to explain the observations made on mothers mosaic for chromosome 21 (Peters et al. 1987).

Using trisomy 19 produced experimentally in the mouse, we tested the hypothesis that relaxed selection, acting after implantation, may explain the maternal age effect observed in humans. Males heterozygous for the Rb(9.19)163H and Rb(5.19)T1Wh translocation chromosomes were crossed to young females (less than six months old) and old females (more than nine months old) genetically unrelated strains of mice (C57BL/6J; DBA/2J; C57BL/10 SnJ). Pregnant litters were collected on day 15 of gestation and standard cytogenetic protocols were used to establish the chromosomal complement of each fetus. No increase (but a decrease) in the observed frequency of trisomy 19 was observed (Table 1).

Table 1. Frequency of trisomy 19 on day 15 in genetically different young and old females mated to the same male heterozygous for the Rb(5.19)T1WH/Rb(9.19)163H translocations

Female	Young		Old	
	Total	Trisomy (%)	Total	Trisomy (%)
C57BL/6J	139	23	107	13
DBA/2J	116	19	64	13
C57BL/10SnJ	98	8	4	0

Our data are concordant with human data, pointing to an absence of age-related maternal age effect acting in translocation Down syndrome. Thus, the weight of evidence does not support the hypothesis of relaxed selection as the basis for the maternal age effect. All the studies, however, have addressed mainly the postrecognition phase of the pregnancy and it might well be that the age-related screening is limited to the prerecognition phase of the pregnancy. Thus, unless the basis of the relaxed selection is different for translocation and standard Down syndrome, it appears that the maternal age effect for standard Down syndrome is due to factors acting at or before implantation, not subsequent to it.

b. Maternal Genotype and Survival of Trisomic Fetuses

Much literature exists in which a wide variety of genotypes and cytogenetic phenomena other than parental gonadal mosaicism are suggested to be associated with the production of a trisomic conception. These include ethnic and racial differences, consanguinity, HLA or alpha-1-antitrypsin genotypes, increased frequencies of nucleolar association between acrocentric chromosomes, the presence of chromosome heteromorphisms (e.g. double-NOR), the presence of aneuploid cells in peripheral blood lymphocytes and the presence of an unrelated balanced chromosome rearrangement (Warburton 1985). In one recent study, a DNA haplotype of chromosome 21 was postulated to be associated with an increased tendency to nondisjunction (Antonarakis et al. 1985). However, the effect of a closely linked gene (linkage disequilibrium) or increased survival of trisomy 21 fetuses with this haplotype could also explain this association.

Thus, using a breeding scheme for the production of trisomic fetuses similar to the one described previously, our laboratory has investigated in detail the possibility that the survival of trisomic fetuses may be under genetic control (Vekemans and Trasler, 1987).

TRISOMY 19

Fetuses with trisomy 19 were produced by crossing males heterozygous for the Rb(9.19)163H and Rb(5.19)T1Wh translocation chromosomes with females of genetically unrelated strains of mice. We observed that the prevalence of living trisomy fetuses on day 15 of gestation varied significantly ($p<0.005$) with the genetic background of the female parent (Table 2). We also observed an inverse relationship between the fetal loss and the observed frequency of trisomy (Figure 3). From these findings, we were able to determine that selective elimination of conceptuses with trisomy 19 was occurring and was under genetic control. A strain of mice with a high frequency of trisomy had a low fetal loss and vice versa.

Table 2. Frequency of Trisomy 19 on Day 15 of Gestation in Different Females Mated to Rb(5.19)T1WH/Rb(9.19)163H Males

Female	Fetuses Examined		Trisomy (%)
	Normal	Trisomic	
C57BL/10SnJ	90	7	7.2
SWV	127	14	9.9
DBA/2J	94	22	19.0
C3H/HeJ	138	34	19.8
C57BL/6J	91	25	21.6

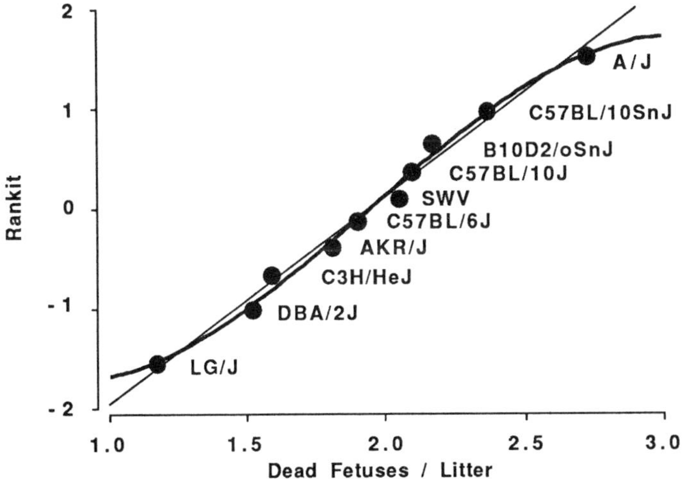

Figure 3. Rankit Analysis of Fetal Loss on Day 15 of Gestation in Females of Genetically Unrelated Strains of Mice

Furthermore, the frequencies of trisomies clustered in two groups; a high group and a low group (Table 2). A rankit analysis of the original and more recent data do indeed suggest they fit a bimodal distribution (Figure 4). This suggests that a very small number of loci (perhaps one) is responsible for the strain difference in the selection process. To determine whether a fetal or maternal locus was responsible for the observed strain difference, reciprocal crosses were performed (Table 3). That is, females heterozygous for the Rb(9.19)163H and Rb(5.19)T1Wh translocation chromosomes were now crossed to males from genetically unrelated strains of mice. No significant difference in the frequency of living trisomy 19 fetuses was detected in the reciprocal crosses. This suggests that a maternal locus expressed perhaps in the uterine environment or contributing to fetal development (maternal imprinting) explains the observed strain difference. In order to determine the target of the maternal locus involved in the selection process, studies were carried out at earlier times of gestation (Vekemans and Jacob 1988).

Preliminary data obtained at the blastocyst stage, and on day 9 of gestation, indicate that the frequency of trisomy 19 embryos is the same in the two genetically unrelated females (Table 4). Therefore, the expression of the maternal locus is taking place presumably after implantation.

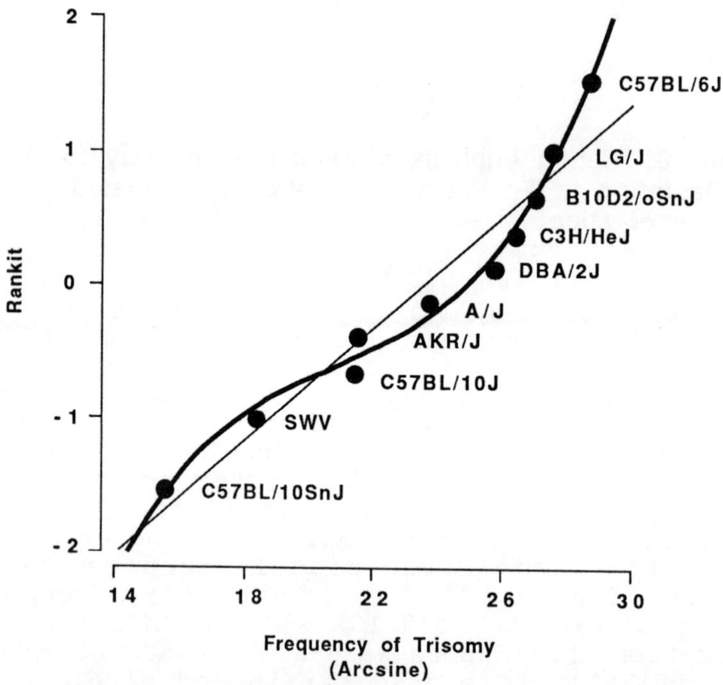

Figure 4. Rankit Analysis of the Frequency of Trisomy 19 on Day 15 of Gestation in Females of Genetically Unrelated Strains of Mice

Table 3. Frequency of Trisomy 19 on Day 15 of Gestation in Rb(5.19)T1WH/Rb(9.19)163H Females Mated to Different Males

Male	Fetuses Examined		Trisomy (%)
	Normal	Trisomic	
SWV	122	18	12.8
DBA/2J	169	30	15.1
C3H/HeJ	110	13	10.6
C57BL/6J	94	11	10.5

Table 4. Frequency of trisomy 19 at different days in gestation in different females mated to Rb(5.19)T1WH/Rb(9.19)163H males

Female	Day 4		Day 9		Day 15	
	Total	Trisomy (%)	Total	Trisomy (%)	Total	Trisomy (%)
SWV	18	50.0	94	14.5	141	9.9
C57BL/6J	15	53.3	56	14.5	116	21.6

TRISOMY 16

In order to determine whether the gene conferring selection is specific for the type of trisomy in question or whether it acts through a common mechanism for all types of trisomy, we also investigated systematically the factors controlling the survival of trisomy 16 fetuses. Trisomy 16 in the mouse has been proposed as a model for trisomy 21 (Down syndrome) in humans because of the homology for at least five genes on mouse chromosome 16 and human chromosome 21. Fetuses with trisomy 16 were produced by crossing males heterozygous for the Rb(6.16)24Lub and Rb(16.17)8Lub translocations with females of several genetically-unrelated strains of mice (Epstein and Vekemans, 1987). Once again, significant differences ($p<0.05$) were detected in the frequencies of trisomy 16 observed in females from genetically-unrelated strains of mice but not in the reciprocal crosses (Table 5).

Table 5. Frequency of Trisomy on Day 15 of Gestation in Different Females Mated to Rb(6.16) 8 Lub/Rb(16.17) 24 Lub Males and in the Reciprocal Crosses

Cross		Fetuses Examined		Trisomy (%)
Female	Male	Normal	Trisomic	
C57BL/6J	Rb(6.16)/(16.17)	127	16	11
SWV	Rb(6.16)/(16.17)	157	39	20
C3H/HeJ	Rb(6.16)/(16.17)	116	31	21
Rb(6.16)/(16.17)	C57BL/6J	132	14	10
Rb(6.16)/(16.17)	SWV	130	23	14
Rb(6.16)/(16.17)	C3H/HeJ	132	21	15

This suggests that a maternal locus also controls the survival of mouse fetuses with trisomy 16. The position on the mouse genome of the maternal gene involved in the selection process will be established by making use of the recombinant inbred strains of mice derived from the strains C3H/HeJ and C57BL/6J which differ significantly in the frequency of trisomy 16. When extrapolated to human (if one makes use of the conservation of syntenic genes between mouse and man), these murine data will be of great value if an attempt is made at identifying in the population, individuals genetically at risk for giving birth to an aneuploid offspring.

c. Maternal Diet and Survival of Trisomic Fetuses

Many factors besides age have been involved in the etiology of aneuploidy. These include extrinsic factors such as radiation, oral contraceptives, spermicides, fertility drugs, alcohol, coffee consumption and seasonality (Kline and Stein, 1985). How these factors influence the observed incidence of aneuploidy is difficult to evaluate since both the conception and the survival of a fetus with a chromosomal anomaly could be affected. For example, smoking at the time of conception was suggested to decrease the risk of trisomic abortion among younger women and increase it among older women (Kline et al. 1983). Two different processes might underlie this association: smoking may alter the probability that a trisomic conception will occur, or smoking may alter the survival of the trisomic conception.

Between April 22, 1986 and May 16, 1986, SWV mice were mistakenly fed Purina Rat Chow instead of Purina Lab Chow. As a result, the frequency of trisomy 19 fetuses went up from 11.2% to 21.6%. This prompted us to investigate the possibility that a relationship might exist between the nutritional status of the mother and the observed frequency of trisomy. Male mice heterozygous for both the Rb(5.19)T1Wh and Rb(9.19)163H translocation chromosomes (for the production of trisomy 19 fetuses) as well as the Rb(6.16)24Lub and Rb(16.17)8Lub translocation (for the production of trisomy 16 fetuses) were mated to female mouse strains C3H/HeJ and SWV. The female mice were fed a diet of either Purina Mouse Chow or Purina Lab Chow prior to mating. A significant reduction ($p<0.05$) in the frequency of trisomy 19 fetuses was detected when C3H/HeJ females were fed Purina Mouse Chow as compared to Purina Lab Chow (Leblanc et al. 1988). Preliminary data indicate that a similar reduction may exist in the frequency of trisomy 16 fetuses when SWV mice are fed Purina Lab Chow. This reduction was not observed with trisomy 16 fetuses in the C3H/HeJ strain (Table 6).

Table 6. Frequency of trisomy 16 and 19 on Day 15 of Gestation in SWV and C3H/HeJ strains of mice fed Lab Chow and Mouse Chow

		Lab Chow	Mouse Chow
	Female	**Trisomy (%)**	**Trisomy (%)**
Trisomy 19	SWV	10	4
	C3H/HeJ	20	10
Trisomy 16	SWV	20	14
	C3H/HeJ	22	18

The finding that a relationship may exist between the frequency of trisomy and the nutritional status of the mother opens the possibility to manipulate the maternal process of elimination of aneuploid fetuses - a process called supportive terathanasia (Warkany, 1978). Further investigations will determine when, during gestation, the dietary-induced selection against trisomic fetuses is occurring.

CONCLUSION

The prevalence of trisomy at a given time of gestation depends on two events: its incidence at conception and the probability of survival of the trisomic fetus up to the time of observation.

The present research has shown in turn that the probability of survival of trisomic fetuses is dependent on several factors. In particular, a high probability of survival of trisomic fetuses is associated with a specific diet or a specific maternal gene, but not increasing maternal age. Future studies are aimed at identifying the specific genetic and environmental factors involved, and understanding their pathogenesis. When extrapolated to humans, it is hoped that such studies will contribute to identifying in the population, individuals at risk for giving birth to an aneuploid offspring and enabling them to benefit from fetal diagnosis and secondary prevention.

ACKNOWLEDGMENTS

The author wishes to thank G. Olynyk and C. Leblanc for their excellent technical assistance. The financial support of the Medical Research Council of Canada is gratefully acknowledged.

REFERENCES

Antonarakis, S., Kittur, S.D., Metaxotou, C., Watkins, P.C., Patel, A.S. (1985) Analysis of DNA haplotypes suggests a genetic predisposition to trisomy 21 associated with DNA sequence on chromosome 21. Proc. Nath. Acad. Sci. USA 82: 3360-3364.

Ayme, S., Lippman-Hand, A. (1982) Maternal age effect in aneuploidy: does altered embryonic selection play a role? Am. J. Hum. Genet. 34: 558-565

Carothers, A.D. (1983) Evidence that altered embryonic selection contributes to maternal age effect in aneuploidy: a spurious conclusion attributable to pooling of heterogenous data. Am. J. Hum. Genet. 35: 1057-1059

Carothers, A.D. (1987) Down syndrome and maternal age: the effect of erroneous assignment of parental origin. Am. J. Hum. Genet. 40: 147-150

Epstein, C.J. (1987) The Mouse Trisomies: Experimental Systems for the Study of Aneuploidy. Issues and Reviews in Teratology. Kalter, H. ed. Vol.3.

Epstein, D.J., Vekemans, M. (1987) Genetic control of the survival of murine trisomy 16. Am. J. Hum. Genet. 41: A346

Gropp, A. (1982) Value of an animal model for trisomy. Virchows Arch. 395: 117-131

Hook, E.B. (1983) Down syndrome rates and relaxed selection at older maternal ages. Am. J. Hum. Genet. 35: 1307-1313

Hook, E.B. (1985)
The impact of aneuploidy upon public health: mortality and morbidity associated with human chromosome abnormalities.
In Aneuploidy: etiology and mechanisms. Eds. V.L. Dellarco, P.E. Voytek, A. Hollander. Basic Life Sciences 36: 7-33

Kline, J., Levin, B., Shrout, P., Stein, Z., Susser, M., Warburton, D. (1983)
Maternal smoking and trisomy among spontaneously aborted conceptions.
Am. J. Hum. Genet. 35: 421-431

Kline, J., Stein, Z. (1985)
Environmental causes of aneuploidy: why so elusive?
In Aneuploidy: etiology and mechanisms. Eds. V.L. Dellarco, P.E. Voytek, A. Hollander
Basic Life Sciences 36: 149-164

Leblanc, C., Olynyk, G., Yelen, S., Trasler, T., Epstein, D.J., Vekemans, M. (1988)
Dietary influence on the survival of murine trisomies.
Genome 30: A32.44.13

Peters, G.B., Ford, J.H., Nicholl, J.K. (1987)
Trisomy 21 mosaicism and maternal age effect.
Lancet i: 1202-1203

Stein, Z., Stein, W., Susser, M. (1986)
Attrition of trisomies as a maternal screening device.
Lancet i: 944-947

Stein, Z., Susser, M., Warburton, D., Wittes, J., Kline, J. (1975)
Spontaneous abortion as a screening device.
Am. J. Epidem. 102: 275-290

Vekemans, M., Jacob, C. (1988)
Controle genetique de la survie des embryons trisomiques chez la souris.
Med. Sc. Suppl. 1: A33

Vekemans, M., Trasler, T. (1987)
Genetic control of the survival of trisomic fetuses in mice: a preliminary report.
Am. J. Med. Genet. 26: 763-770

Warburton, D. (1985)
Genetic factors influencing aneuploidy frequency.
In Aneuploidy: etiology and mechanisms
Eds. V.L. Dellarco, P.E. Voytek, A. Hollander
Basic Life Sciences 36: 133-148

Warkany, J. (1978)
Terathanasia
Teratology 17: 187-192

Molecular and Cytogenetic Studies of
Non-Disjunction, pages 263–275

SOME REFLECTIONS ON NONDISJUNCTION IN DROSOPHILA

DAN L. LINDSLEY

Department of Biology, University of California, San Diego, La Jolla, California 92092

In 1981 a symposium was held entitled "Trisomy 21--Research Perspectives". In that symposium my late good friend and closest scientific colleague, Larry Sandler, presented a paper entitled "The Meiotic Nondisjunction of Homologous Chromosomes in Drosophila Females". Any similarity between this paper and his is purely intentional. I will open this discussion with a historical look at the early work of Calvin B. Bridges on nondisjunction in females and follow with more recent insights into the mechanism of nondisjunction, both spontaneous and X-ray induced. I will close the discussion with consideration of nondisjunction of the *X* and *Y* in males and some of the surprising correlated phenomena.

X-CHROMOSOME NONDISJUNCTION IN FEMALES

Nondisjunction was first described by Bridges (1913, 1914, and 1916) in the second decade of this century. It is instructive to view his results and conclusions in terms of the understanding of the day rather than of what is known today. As we now know, during the first meiotic division in females, *X* chromosomes nondisjoin roughly one time per five hundred to one thousand meioses; the diplo-*X* and nullo-*X* ova are recovered as *XXY* females and *X0* males when fertilized by *Y*-and *X*-bearing sperm respectively. This observation that the *Y* chromosome is without effect on the sexual phenotype of these exceptional offspring surely caused a significant delay in recognition of the male-determining nature of the mammalian *Y* chromosome. The exceptional *XO* males are normal in phenotype and behavior; they are, however, sterile. The *XXY* females are normal in phenotype, behavior and fertility; they are distinguished only by the fact that they produce about 5% so-called secondary exceptional progeny (*XXY* matroclinous daughters and *XY* patroclinous sons) as opposed to 0.1% primary exceptions from *XX* females.

In his first paper defining nondisjunction, Bridges (1913 at the age of 24) describes a progeny with 5% sex-chromosome exceptions; both from the frequency of exceptional progeny and the fact that the patroclinous sons were fertile, it is clear that the parental female was *XXY* and that Bridges was observing secondary nondisjunction. However, seventy-five years ago males of *Drosophila ampelophila* (as *D. melanogaster* was known at the time) were considered, on the basis of earlier cytological observations of Stevens (1908), to carry an *X* chromosome but no *X* homologue. Accordingly, the ability of exceptional females to in turn produce high frequencies of exceptional offspring was attributed to homozygosity for an *X*-linked recessive gene, said homozygosity being passed from mother to matroclinous daughter.

Between 1913 and 1914 Bridges made two important further observations. First, he noted that not only exceptional daughters, but also a fraction of the regular daughters of matroclinous females displayed high rates of nondisjunction as did many of the daughters of their regular sons. This observation together with his inability to map a gene responsible for increased nondisjunction led him to abandon the mutant gene hypothesis. The second important observation was his cytological discovery of the presence of the *Y* chromosome as the normal homologue of the *X* chromosome in males and his demonstration that females producing high levels of nondisjunctional progeny were *XXY*. From these observations he arrived at the model for secondary nondisjunction set forth in the second paragraph. As expected, the regular *XXY* daughters of matroclinous females carry one *X* chromosome from each parent and are therefore frequently heterozygous. From his observation that the secondary exceptional daughters of such females are heterozygous in the same way as their mothers, Bridges concluded that nondisjunction occurs at the reductional division. Perhaps his most fundamental conclusion in the 1914 paper is the following: "*there can be no doubt that the complete parallelism between the unique behavior of the chromosomes and the behavior of the sex-linked genes and sex in this case means that sex-linked genes are located in and borne by the* X *chromosome.*"

His 1916 paper, a 108 page work entitled "Nondisjunction as proof of the chromosome theory of heredity" published in two parts beginning on page 1 of volume 1 of Genetics, represents the culmination of Bridges' work on nondisjunction. In it he explicitly recognizes and defines the difference between primary nondisjunction of the *X* chromosomes in *XX* females and secondary nondisjunction of the *X*'s in *XXY* females.In that paper, Bridges first encounters equational exceptions, *i.e.* exceptional females homozygous for alleles for which their mothers were heterozygous. Upon test crossing he was able to show that such exceptions carried one recombinant and one non-recombinant chromosome; *e.g.*, a w^e exceptional daughter of a w^e $v/+$ $+$ mother could be shown to be w^e v/w^e $+$ in genotype. Until that time it had generally been assumed that crossing over took place in the two-strand stage, but Bridges recognized that this result demanded recombination in the four-strand stage in accordance with the chiasmatype hypothesis, which had been proposed earlier by Janssens (1909) based on cytological observations of spermatogenesis in a salamander.Bridges noted that wherever it was possible to determine, equational exceptions carried one recombinant and one nonrecombinant chromosome. At this point there seems to have been a lapse in his logic. He believed, as illustrated in the diagram reproduced in Figure 1, that in order to achieve homozygosis nondisjunction must have occurred at the equational division; he infers from the regularity with which equational exceptions are associated with crossing over that there is some causal relation between exchange in the four-strand stage and equational

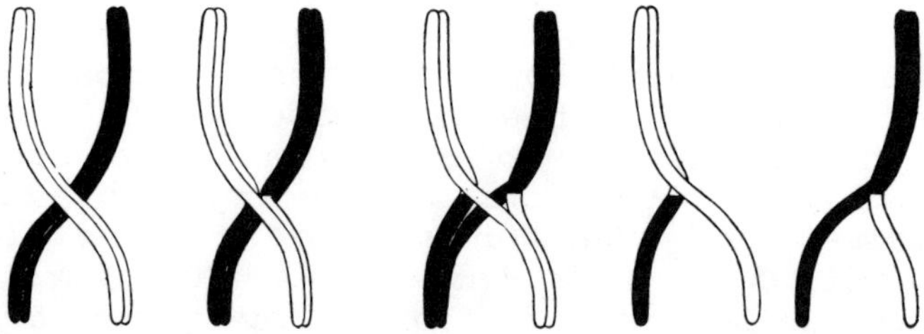

FIGURE 1. Bridges' (1916) figure illustrating his rationale for postulating equational nondisjunction (after Morgan, Sturtevant, Muller, and Bridges, 1915)

nondisjunction. He apparently did not explore the consequences of reductional nondisjunction following exchange at the four-strand stage. He also did not seem to be prepared to accept the notion of four-strand crossing over as the normal state of affairs rather than a rare event associated with the generation of equational exceptions. His data show that homozygosis is seen for loci at the left end of the genetic map of the X more often that at the right end. We realize today that such equational exceptions indicate reductional nondisjunction following crossing over between the loci becoming homozygous and the centromere. I once asked Sturtevant why this was not obvious to Bridges; he replied that the position of the centromere on the genetic map was unknown at the time and that, based on his supposition that equational exceptions represented products of nondisjunction at the equational division, Bridges had in fact inferred the location of the centromere to be at the left end (rather than at the right end where we now know it to reside) . However, examination of Figure 1 reveals that the centromere is not a feature of Bridges' representation of the chromosomes in the four-strand stage, nor is there any indication that the role of the centromeres in determining the disjunctional behavior of the chromosomes, was considered at all in his analysis of the problem.

THE RELATION BETWEEN EXCHANGE AND DISJUNCTION

Bridges attributed secondary nondisjunction to heterosynapsis between one *X* and the *Y* chromosome with the other *X* behaving as a univalent and segregating at random. A corollary of this hypothesis is that the *X* chromosomes of secondary exceptional females are expected to be nonrecombinant. Because he eliminated equational exceptions from consideration, Bridges found that this expectation was realized. The inclusion of equational exceptions in the calculations leads to a crude estimate from Bridges" data that 2.6% of secondary exceptions are derived from tetrads that have undergone exchange. Genotypic analysis of a sample of 140 exceptions derived from well marked *XXY* mothers by Carpenter (1973) led to an estimate that the population of tetrads contributing to secondary exceptions comprised 79.1% no-exchange (E_0), 17.5% one-exchange (E_1) and 3.4% two-exchange (E_2) tetrads; it is difficult to reconcile these observations with those of Bridges. The contribution of exchange tetrads not withstanding, it is nevertheless the case that secondary exceptions are preferentially recovered from no-exchange tetrads. In their study of the meiotic behavior of inversion heterozygotes, Sturtevant and Beadle (1936) showed that the increased incidence of *X*-chromosome no-exchange tetrads brought about by inversion heterozygosity leads to a commensurate increase in exceptional offspring in XXY females. *XX* females heterozygous for the same inversions, despite having the same elevated frequencies of no-exchange tetrads, display regular *X*-chromosome segregation. This observation suggests the existence of a fail-safe mechanism that assures the regular segregation of homologues which are not conjoined by chiasmata, but which in the case of the *X* chromosomes is upset by the presence of a *Y* chromosome. These observations are rationalized by the distributive pairing hypothesis so ably propounded by Rhoda Grell (1976) and which will be discussed by Scott Hawley in the next paper.

PRIMARY NONDISJUNCTION

Data on the rate of primary nondisjunction of both the *X* and the fourth chromosomes in

Drosophila females were collected as the control in experiments designed to examine X-ray-induced nondisjunction (Parker *et al.*, 1974); they are presented below in terms of the constitutions of the ova producing the progeny scored. Both sex-chromosome and fourth-chromosome exceptions are recovered in approximately one case per thousand

	mono 4	diplo 4	nullo 4
mono X	54,508	28	20
diplo X	25	0	1
nullo X	24	0	4

offspring; however, the relative recovery of diplo-*X* and nullo-*X* ova is 50% that of mono-X ova, whereas fourth-chromosome exceptions are recovered in proportion to their incidence. Apparently, nondisjunction of *X*'s and *4*'s are not independent; double exceptions are recovered about one hundred times more frequently than expected based on independence.

Owing to the rarity of primary exceptions, there are few data on the relation between exchange and primary nondisjunction. The single work by Merriam and Frost (1964) involves the genetic analysis of 149 primary exceptions recovered among progeny of multiply marked *XX* females. By test crossing, they were able to ascertain the genotypes of these half tetrads and determine that tetrads giving rise to nondisjunction are not representative of the overall tetrad population. The distribution of no-, one-, two-, and three-exchange tetrads giving rise to nondisjunction is as follows: $E_0 = 0.2597$, $E_1 = 0.2453$, $E_2 = 0.4755$, and $E_3 = 0.0195$ compared to $E_0 = 0.046$, $E_1 = 0.657$, $E_2 = 0.287$, and $E_3 = 0.010$.derived from their regular sisters This is apparently brought about by a 3.25-fold enrichment of no-exchange tetrads and a four-fold reduction in the contribution of one-exchange tetrads in the nondisjunctional population. Although one can imagine reasons why no-exchange tetrads are more likely to nondisjoin than others, it is not clear why one-exchange tetrads appear to segregate so much more regularly that those with two or more exchanges. It has been suggested (e.g. Merriam and Frost, 1964) that homologous chromosomes in multiple-exchange tetrads occasionally have difficulty in disengaging and accordingly become included in the same telophase I nucleus.

X-RAY-INDUCED NONDISJUNCTION IN FEMALES

One of the first observed genetic effects of X irradiation in Drosophila was the induction of nondisjunction of the sex chromosomes in females (*e.g.*, Mavor, 1924). Muller's (1927) discovery of the mutagenic and clastogenic effects of X rays, however, overshadowed the disjunctional effects and they were largely neglected until a serious analysis was undertaken by Dean Parker in the late 1960's. He postulated that the observed nondisjunction is a by-product of induced interchange between nonhomologous chromosomes in the primary oocyte, and he and Williamson (1970) constructed a

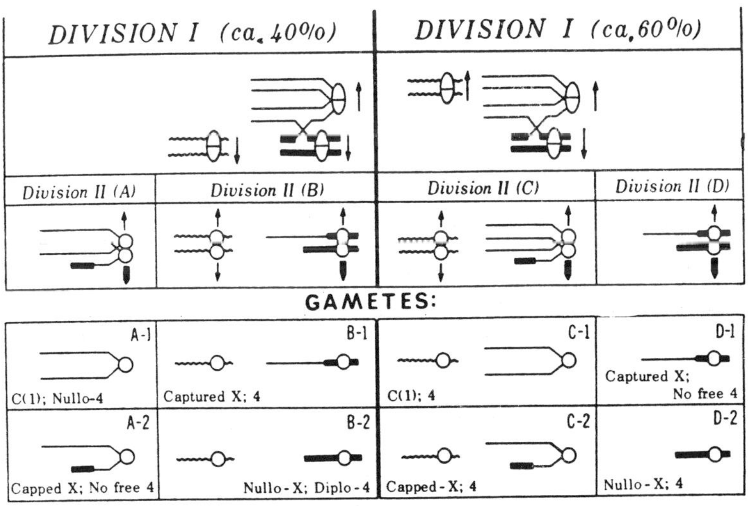

Figure 2, Meiotic segregation from the pseudo-bivalent formed by induced exchange between an attached *X* and a fourth chromosome, with the uninvolved fourth chromosome segregating at random. (Parker and Williamson, 1970).

genotype to test the proposition. In this particular instance nondisjunction of the minute fourth chromosomes was scored in a situation in which X-ray-induced interchange between one of the fourth chromosomes and attached-*X* chromosomes was a frequent occurrence. If, in an interchange oocyte, as Parker postulated, the participating attached *X*'s and fourth chromosome form a pseudo-bivalent that segregates regularly, allowing the uninvolved fourth chromosome to segregate at random, then nullo-*X*, diplo-*4* ova will be produced as one of the products of meiosis (Fig.2, B2). Upon fertilization by an *X*-bearing sperm, such ova produce triplo-four males; these males are expected to carry homologous fourth chromosomes from their mothers and a single fourth chromosome from their fathers. The reciprocal product, an attached-*X*, nullo-*4* ovum (Fig. 2, A1), upon fertilization by a *Y*-bearing sperm produces poorly viable haplo-*4* daughters, whose frequency cannot be reliably determined. Fourth-chromosome nondisjunction unassociated with attached-*X*-four interchange is expected to produce triplo-four daughters and sons in equal numbers. Parker and Williamson observed 269 triplo-four sons and 9 triplo-four daughters; interestingly, all eleven triplo-four offspring observed in the unirradiated control crosses were males.

As further indication that fourth-chromosome nondisjunction is associated with chromosome rearrangement they demonstrated that the incidence of triplo four exceptions increases exponentially with increasing dose of X rays (proportional to the 1.965 power of the dose, equivalent to detachment which increases as the 1.755 power in the experiments described); furthermore, the incidence of triplo four offspring parallels precisely that of detachment in broods of eggs laid on successive days following exposure to X rays. The experimental situation described above as well as another set of experiments designed to examine the consequences of pseudo-bivalents involving the *Y* and fourth chromosomes (Parker, 1970), lead to relatively clear-cut results, since the pseudo-bivalents induced involve chromosomes which are not conjoined to their homologues by chiasmata (*i.e.* the fours and the *X* and *Y*). Experiments designed to examine interactions between free-*X* chromosomes, which are usually conjoined by chiasmata, and the fourth chromosomes are more complicated and the results more difficult to interpret; nevertheless, Parker *et al.* (1974) conclude that they support the hypothesis that X-ray-induced nondisjunction is causally related to induced interchange between nonhomologous chromosomes.

NONDISJUNCTION OF SEX CHROMOSOMES IN MALES

Despite the fact that homologous chromosomes are not conjoined by chiasmata in primary spermatocytes, segregation is highly regular in Drosophila males; the rates of nondisjunction of the sex and the fourth chromosomes are indicated in the accompanying data on sperm constitutions of chromosomally normal males collected by Ivy (1981).

	mono 4	diplo 4	nullo 4
X	2984	3	0
Y	2723	3	0
XY	2	0	0
0	3	0	0

The *X-Y* bivalent as viewed at first meiotic metaphase involves an association between the short arm of the submetacentric *Y* chromosome and the proximal heterochromatin of the *X*. A deficiency for most of the *X* heterochromatin (designated *Xh-*) is unable to pair with the *Y* and accordingly the two chromosomes segregate as univalents. Cytological studies (McKee and Lindsley, 1987) indicate that meiosis in the most extreme *Xh-* males generates equal numbers of *X*, *Y*, *XY*, and nullo-*X*, nullo-*Y* spermatids. However, these genotypes are not recovered equally among the progeny of such males, *e.g.*, a typical progeny of males carrying $In(1)sc^{4L}sc^{8R}$, such a proximal deficiency, comprises 0.33 *XX*, 0.16 *XY*, 0.05 *XXY*, and 0.46 *X0*. The skewing of the ratios, termed meiotic drive, is demonstrably not caused by zygote mortality but is attributable to sperm dysfunction (Peacock *et al.*, 1975; McKee, 1984).

McKee has provided evidence that the probability of function of a sperm produced by *Xh-/Y* males is inversely related to its chromosome content such that each chromosome arm acting independently reduces recovery, the recovery of *X* being 45% that of no *X* and *Y* 25% that of no *Y*, and that of *XY* being 0.45 X 0.25 = 0.11 that of neither. By utilizing a *2-3* translocation in which the entire left arm of chromosome *3* is translocated to the end of the right arm of chromosome *2*, McKee was able to show that each autosomal arm reduces sperm recovery by approximately 50% in *Xh-/Y* males; he examined the recovery of sperm with three, four, and five autosomal arms produced as segregants from heterozygotes for the above translocation; in *X/Y* males reciprocal meiotic products were recovered equally frequently, whereas in *Xh-/Y* males five-arm-bearing sperm were recovered with but one fourth [$(1/2)^2$] the frequency of three-arm sperm, and this ratio was independent of the sex-chromosome constitution of the sperm. Thus, the autosomal complement reduces sperm recovery to $(1/2)^4$ = 0.0625; accordingly, the expected frequencies of functional sperm in *Xh-/Y* males are 0.0625 X 0.45 = 0.028 *X*, 0.625 X 0.25 = 0.016 *Y*, 0.625 X 0.10 = 0.00625 *XY*, and 0.0625 X 1.0 = 0.0625 nullo *X*, nullo *Y* for an overall sperm viability of approximately 2.8%. It might be added that Peacock *et al.* (1975) detect considerable ultrastructural damage during spermiogenesis in *Xh/Y* males, but not to an extent commensurate with 2.8% functionality of sperm; furthermore, despite such a low estimated frequency of functional sperm, *Xh-/Y* males are fertile. However, males carrying *Xh-* in combination with either a *Y*-autosome translocation (Lindsley, *et al.*) or a *Y* chromosome with a substantial portion of *X*-derived heterochromatin appended (Rahman and Lindsley, 1981) are sterile.

In an experiment in which a number of proximally deficient *X* chromosomes were examined for their effects on nondisjunction, meiotic drive, and male fertility in combination with *T(Y;A)* and *Dp(X;Y)*, McKee (McKee and Lindsley, 1987) found that all of these phenotypes are highly correlated; deficiencies that disjoin regularly from the *Y* display equal recovery of reciprocal meiotic products and are fertile in combination with *T(Y;A)*'s and *Dp(X;Y)*'s; those characterized by varying frequencies of nondisjunction show proportional levels of meiotic drive and are sterile in the indicated genotypes. In his attempt to map the heterochromatic region responsible for the behavior described, McKee noted that segregation was normal in the presence of as few as six to eight of the 250 ribosomal cistrons normally present, whereas loss of the entire ribosomal DNA complement was associated with nondisjunction, meiotic drive, and male infertility. His suspicion that the ribosomal cistrons are responsible for X-Y pairing and that such pairing is crucial to normal spermiogenesis received powerful support from his recent finding (McKee, 1989)that a single ribosomal DNA transcription unit comprising both 18S and 28S sequences, when integrated by means of a P-element transpostion into the distal euchromatin of the *X*, can rescue the effects of proximal heterochromatic deficiencies. For example, recombining a single ribosomal transcription unit onto the distal end of an *X* chromosome that carries a large proximal heterochromatic deficiency, which can be shown (a) cytologically not to pair with a Y chromosome, (b) to segregate independently of the *Y*, (c) to exhibit high levels of meiotic drive, and (d) to be male sterile in combination with *Dp(X,Y)*, can partially restore *X-Y* pairing, *X-Y* disjunction, normal segregation ratios, and male fertility in the presence of *Dp(X;Y)*. Two doses of the rDNA insert are more effective in rescue than one.

It is perhaps interesting to note that pairing between the nucleolus organizing region of chromosome *21* with similar regions on nonhomologous chromosomes has been invoked as contributory to chromosome-*21* nondisjunction in humans (Babu and Verma, 1985)

REFERENCES

Babu KA, Verma RS (1985). Structural and functional aspects of nucleolus organizer regions (NORs) of human chromosomes. Int Rev Cytol 94:151-176.

Bridges CB (1913) Non-disjunction of the sex chromosomes of Drosophila. J Exp Zool 15:587-606.

Bridges CB (1914) Direct proof through non-disjunction that the sex-linked genes of Drosophila are borne by the X-chromosome. Science 40:107-109.

Bridges CB (1916) Non-disjunction as proof of the chromosome theory of heredity. Genetics 1:1-57; 107-163

Carpenter ATC (1973). A meiotic mutant defective in distributive disjunction in *Drosophila melanogaster*. Genetics 73:393-428.

Grell RF (1976). Distributive pairing. In Ashburner M, Novitski E (eds.) "The Genetics and Biology of Drosophila".New York: Academic Press, vol 1a, pp 435-486.

Ivy JM (1981). Mutations that disrupt meiosis in males of *Drosophila melanogaster*. PhD thesis, University of California, San Diego. pp 1-93.

Janssens FA (1909). La theorie de la chiasmatypie. La Cellule 25:389-414.

Lindsley DL, Pearson C, Rokop SR, Jones M, Stern D (1979). Genotypic features causing sterility of males carrying a bobbed-deficient *X* chromosome: translocations involving the *Y* chromosome and chromosome *2* or *3*. Genetics 91:569 (abstract).

Mavor JW (1924). The production of non-disjunction by X-rays. J Exp Zool 39:381-432.

McKee B (1984). Sex chromosome meiotic drive in *Drosophila melanogaster* males. Genetics 106: 403-422.

McKee B (1989). Meiotic drive, sex-chromosome pairing, and ribosomal DNA in Drosophila males. Am Nat (in press).

McKee B, Lindsley DL (1987). Inseparability of *X*-heterochromatic functions responsible for *X:Y* pairing, meiotic drive, and male fertility in *Drosophila melanogaster*. Genetics 116:399-407.

Merriam JR, Frost JN (1964). Exchange and nondisjunction of the *X*-chromosome in female *Drosophila melanogaster*. Genetics 49:109-122.

Morgan TH, Sturtevant AH, Muller HJ, Bridges CB (1915). "Mechanism of Mendelian Heredity." New York: Henry Holt, p 133.

Muller HJ (1927). Artificial transmutation of the gene. Science 66:84-87.

Parker DR (1970). Coordinate nondisjunction of Y and fourth chromosomes in irradiated compound-X females. Mutat Res 9:307-322.

Parker DR, Williamson JH (1970). Heterologous interchange at meiosis in Drosophila III. Interchange-mediated nondisjunction. Mutat Res 9: 273-286.

Parker DR, Williamson JH, Gavin J (1974). The nature and time of occurrence of radiation-induced nondisjunction of the acrocentric X and fourth chromosomes in *Drosophila melanogaster* females. Mutat Res 24:135-148.

Peacock WJ, Miklos GLG, Goodchild DJ (1975). Sex chromosome meiotic drive systems in *Drosophila melanogaster* I. abnormal spermatid development in males with a heterochromatin-deficient *X* chromosome (sc^4sc^8). Genetics 79:613-634.

Rahman RM, Lindsley DL (1981). Male-sterilizing interactions between duplications and deficiencies for proximal *X*-chromosome material in *Drosophila melanogaster*. Genetics 99:49-64.

Sandler LM (1981). The meiotic nondisjunction of homologous chromosomes in Drosophila females. In de la Cruz FF and Gerald PS (eds): "Trisomy 21 (Down syndrome) research perspectives" Baltimore: University Park Press, pp 189-197.

Stevens, NM (1908). A study of the germ cells of certain Diptera, etc. J Exp Zool 5:359-374.

Sturtevant AH, Beadle GW (1936) The relations of inversions in the X chromosome of Drosophila melanogaster to crossing over and disjunction. Genetics 21:554-604.

Molecular and Cytogenetic Studies of
Non-Disjunction, pages 277–302

GENETIC AND MOLECULAR ANALYSIS OF A SIMPLE DISJUNCTIONAL SYSTEM IN DROSOPHILA MELANOGASTER

R. Scott Hawley

Department of Molecular Genetics, Albert Einstein College of Medicine, 1300 Morris Park Avenue, Bronx, NY 10461

SUMMARY

The basic aspects of the achiasmate distributive segregation system in Drosophila are briefly reviewed. Specific attention is given to the role of homology in specifying the patterns of distributive segregation. The genetic and molecular analysis of mutations affecting the system is also discussed. Finally, a consideration of the available evidence allows the presentation of a revised model of meiosis in Drosophila.

INTRODUCTION

In most eukaryotes, segregation of homologues at meiosis I is facilitated by homologous pairing and exchange, followed by the disjunction of chiasmate bivalents at anaphase I. Indeed, the vast body of evidence (reviewed by Hawley, 1988) leads to the conclusion that normal levels of exchange are both necessary and sufficient to ensure disjunction.

Nonetheless, in a number of organisms, a growing body of evidence has demonstrated the existence of a secondary, or back-up, disjunctional system which guarantees the segregation of those chromosomes that have failed to undergo exchange. Such an exchange independent disjunction system has been identified in Drosophila melanogaster females and is referred to as "distributive segregation".

Distributive segregation and exchange-mediated disjunction have been argued to be a separate meiotic processes both in terms of their meiotic properties, and in terms of meiotic mutations whose effects are apparently specific to one system or the other. Although distributive disjunctions are as efficient and reliable as exchange-mediated disjunctions, the system seems to be dispensable. One can easily recover and analyze mutations which ablate various aspects of this process at the genetic, cytological, and biochemical levels.

This manuscript will attempt to summarize the results of such analyses by ourselves and others. In addition, a simple, but speculative, modification of a model originally proposed by Novitski (1964) to explain the function of this system in meiosis will be presented.

THE DISCOVERY OF THE DISTRIBUTIVE SYSTEM

Although exchange itself is sufficient to ensure proper chromosome segregation in Drosophila melanogaster females, several lines of evidence demonstrate the existence of a second, exchange-independent, system of segregation. In females with structurally normal X chromosomes, the frequency of nonexchange bivalents is approximately 6.0% while the frequency of spontaneous X nondisjunction is approximately 0.5% (Bridges, 1916; and Weinstein, 1936). Although one might have expected that inversion heterozygotes, which reduce the frequency of exchange between X chromosomes, would show elevated levels of nondisjunction, these females exhibit little more disjunction than is observed in control females. This result demonstrates that the exchange-mediated system cannot be the only disjunctional mechanism present in Drosophila females. Finally, although the fourth chromosomes in Drosophila are always nonexchange, they segregate from each other with great efficiency (the standard frequency of fourth chromosome nondisjunction is approximately 0.1%). Again, these data demonstrate the existence of an exchange-independent mechanism of segregation.

Bridges (1916) obtained evidence that such exchange-independent segregations followed a different set of rules than did disjunctions mediated by exchange. When examining sex chromosome disjunction in *XXY* Drosophila females, he observed frequencies of *X* nondisjunction much higher than those observed in *XX* females. These *X* nondisjunctional events, termed secondary nondisjunction, involved nonexchange *X* chromosomes that had segregated from the *Y* chromosome (*XX*<->*Y*). When the *X* chromosomes underwent exchange, they segregated from each other faithfully with the *Y* going at random. Thus, the segregational behavior of nonexchange *X* chromosomes was distinctly different from that exhibited by *X* chromosomes which had undergone exchange.

Grell (1962a and 1962b) first proposed the existence of a novel segregation system, termed distributive segregation, to explain both secondary nondisjunction and the disjunction of the always nonexchange fourth chromosomes. Grell demonstrated that the province of the distributive system was limited to nonexchange chromosomes. According to Grell's model, the presence of a chiasma precludes both members of a bivalent from entering into distributive associations.

Through a series of clever and elegant experiments, Grell showed that once nonexchange chromosomes enter the distributive system, the primary criterion for partner choice is availability (i.e. when only two nonexchange chromosomes are present in the same meiotic cell, they will faithfully disjoin from each other regardless of their identity). This is true even when the chromosomes involved are nonhomologous or greatly different in terms of size or shape (i.e. centromere position). For example, in females bearing attached-*X* and attached-fourth chromosomes (both of which are obligate denizens of the distributive system), the two attached chromosomes faithfully segregate from each other despite at least a ten to twenty-fold difference in size. Grell (1962a and 1976) further argued that when three or more chromosomes are available, the choice of disjunctional partners is made on the basis of similarities in chromosome size and/or shape.

An example of the effect of chromosomal size on the choice of disjunctional partners is provided by the meiotic behavior of small *X* chromosomal fragments (free duplications). These chromosomes are usually quite small, relative to a normal *X*. In addition, they do not undergo exchange, and thus their disjunction is facilated by the distributive system. In *XX* females bearing such a duplication, which are otherwise genetically normal, one can measure the ability of the duplication to interfere with the segregation of the always achiasmate fourth chromosomes. In experiments performed by O'Tousa (1982), the ability of a given duplication to induce the nondisjunction of the fourth chromosomes is primarily determined by the size of the duplication (see Figure 1). Those duplications which are similar in size to chromosome four induce high levels of nondisjunction. As diagrammed below, these nondisjunctions result from improper partner recognition such that one fourth chromosome segregates from the duplication while its homolog (the other fourth chromosome) segregates at random.

Duplication <---> 4
<-4->

Duplications which are substantially larger or smaller than a normal fourth chromosome are less effective at interfering with fourth chromosome partner choice.

The role of shape in determining distributive disjunctions is demonstrated by secondary nondisjunction. In general, a number of studies have shown that when two acrocentric and one metacentric chromosome (like the two X chromosomes and a Y) are nonexchange, the metacentric chromosome will disjoin efficiently from the two acrocentric chromosomes (for review see Grell, 1976). Based on a model proposed by Cooper (1948), it is believed that such segregations result from the formation of a trivalent in which each arm of the metacentric is associated with one of the acrocentric chromosomes.

These and other experiments have led to a model in which distributive segregations are entirely independent of homology (Grell, 1976). Several studies have suggested, however, that the parameters of size and shape are insufficient to fully explain partner choice, and suggest a possible role of homology in distributive disjunctions (c.f. Novitski and Puro, 1978). For example, Gershenson (1940) and Lindsley and Sandler (1958) observed that in females heterozygous for an X chromosome inversion which suppresses exchange, small X chromosomal free duplications were capable of inducing significant frequencies of X nondisjunction. Indeed, Anne Zitron and I have shown that in females in which X chromosomal exchange is strongly suppressed by inversion heterozygosity, a small free X duplication (approximately twice the size of chromosome 4) induces more than twice as many XX<->duplication disjunctions as it does 44<->duplication disjunctions. Given that the duplication is much closer in size to chromosome 4 than to the X, these data are not consistent with the size rule. Since all three chromosomes are acrocentric, the disjunctional patterns are also

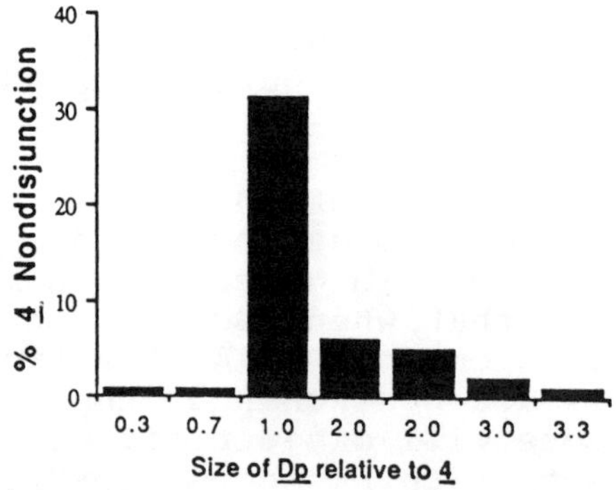

Figure 1: The Effect of a Free Duplication on 4 Nondisjunction (O'Tousa, 1982)

not explicable by the shape rule. All three of these examples suggest that the X and the duplication chromosomes recognize each other by virtue of their relatedness, even in an instance where the homology present is insufficient to allow exchange.

We argue therefore, that under competitive situations, homolog recognition may play some role in the choice of partners within the distributive system. Moreover, as will be discussed in detail below, Anne Zitron and I have recently analyzed a new meiotic mutation, named Axs, which is specifically defective in the recognition of homology within the distributive system.

Taken together these data argue that the choice of disjunctional partners in the distributive system, while homology-independent under noncompetitive situations, is based on factors such as size, shape, and homology when more than two elements are present.

The demonstration of homology-dependent distributive associations allows us to envision the following revised model of meiosis in

Drosophila females. In early prophase homologous chromosome pairing provides an opportunity both for exchange between homologs and for distributive associations. As mentioned above, exchange and subsequent chiasma formation, allow for proper disjunction at anaphase I. The failure to undergo exchange, due either to chance or the presence of structural heterozygosity, does not automatically result in high frequencies of nondisjunction. Rather, achiasmate chromosomes are dealt with by the distributive system which will ensure their proper segregation, albeit by a different set of rules.

There exist two major schools of thought in terms of the timing of pairing for exchange versus the timing of partner choice within the distributive system. According to the first model, distributive associations are made following homologous pairing and exchange; indeed this model argues for a second round of chromosome pairing (Grell 1962a, but for review see Grell, 1976). According to the second model, known as the chromocentral hypothesis, distributive associations occur in the meiotic chromocenter. The term chromocenter is used to describe a "bouquet" arrangement of chromosomes in which the centromeres and heterochromatic regions are joined at one central region (Oksala, 1958). These associations are said to occur either prior to or concurrently with pairing for exchange, and then subsequently voided or over-ridden by the formation of chiasmata. However, according to this model, if exchange does not occur these chromocentral associations will then ensure disjunction (for review see Novitski and Puro, 1978).

We will return to a consideration of these two models after a discussion of the genetic and cytological studies of the distributive system.

THE MUTATIONAL ANALYSIS OF DISTRIBUTIVE SEGREGATION

Based on models proposed by Carpenter (1973) and O'Tousa (1982), distributive segregation can be formally divided into three phases. The first phase is the determination and selection of nonexchange chromosomes, the second phase is the choice and orientation of chromosome partners, and the third phase is the actual disjunction of chromosomes to opposite poles. Although many mutations affecting the frequency and distribution of exchange have been isolated and characterized (for a review see Baker and Hall, 1976), to date only four mutations affecting the distributive system have been isolated; mei-S51, nod, ald, and Axs. Each phase of distributive segregation has been genetically defined by these mutations.

The first phase of the model, namely, the selection of nonexchange chromosomes, is defined by the ald (altered disjunction) mutation (O'Tousa, 1982), which allows exchange chromosomes to enter the system and participate in nonhomologous disjunctions such as XX<->Y and XX<->44. Because the ald mutation allows exchange chromosomes to enter the distributive system, the locus may define a component required for chiasma maintenance (see Hawley, 1988). In addition, ald exhibits an effect on the proper choice of disjunction partners (see below).

As noted above, partner choice within the distributive system is dependent on the recognition of various parameters of chromosome structure, such as size, shape, and possibly, homology. Mutations affecting partner choice are distinguished by phenotypes which include high frequencies of nonhomologous XX<->44 disjunctions and/or improper size recognition.

The first mutation shown to affect partner choice within the distributive system was mei-S51 (Robbins, 1971), a synthetic mutation composed of two recessive genes present on the second and third chromosomes. Females homozygous for mei-S51 exhibit both reduced exchange and high frequencies of nonhomologous disjunction, particularly with respect to the *X* and fourth chromosomes (i.e. *XX*<->*44* segregations are frequent). Robbins proposed that *mei-S51* disrupts a number of aspects of chromosome pairing and alignment prior to metaphase, and therefore, both reduces exchange and prevents proper partner choice within the distributive system. In this sense *mei-S51* may be thought to encode a common component of the mechanisms which ensure homologue pairing for exchange and which determine partner choice within the distributive system.

The *ald* mutation also alters the choice of segregational partners within the distributive system, and thus, induces high frequencies of nonhomologous disjunctions involving the *X* and fourth chromosomes (i.e. *XX*<->*44*). Moreover, *X* chromosomal free duplications which are either larger or smaller than the fourth chromosome induce higher frequencies of fourth chromosomal nondisjunction in *ald* females than are observed in control females. Thus, the *XX*<->*44* nonhomologous segregations mentioned above may simply reflect a more general defect in size recognition in *ald* females.

As noted above, Anne Zitron and I have recently analyzed a new dominant mutation *Axs* (*Aberrant X segregation*) whose effect is specific to the distributive system (Zitron and Hawley, unpublished). *Axs* exhibits at best a weak effect on the frequency of *X* chromosomal exchange (see below), and no effect on the disjunction of chiasmate bivalents. However, when the *X* chromosomes fail to undergo exchange, *Axs* induces high frequencies of *X* nondisjunction, and also increases frequency of fourth chromosome nondisjunction in response to the presence of nonexchange *X* chromosomes. In *XXY* females *Axs* strongly decreases the frequency of *XX*<->*Y* trivalent formation and thus, interferes with *XX*<->*Y* secondary nondisjunction. In most cases, the improper disjunction of *X* and/or fourth chromosomes is associated with concommitment nonhomologous disjunctions (such as *XX*<->*44* or *XXY*<->*44*), rather than random *X* or fourth chromosome disjunction. *Axs* thus decreases the frequency of homolog-homolog segregations within the distributive system, while increasing the frequency of disjunctions involving nonhomologs. We argue that *Axs* defines a locus required primarily for homologous disjunctions, and further reveals the existence of a homology dependent component of the distributive system.

The third and final phase of distributive segregation is the proper segregation of oriented chromosomal partners to opposite poles and is defined by the *nod* (*no distributive disjunction*) mutation (Carpenter, 1973). This locus is required for the actual disjunction of chromosomes within the distributive system. Mutations at the *nod* locus have no effect either on the frequency of exchange or on the disjunction of bivalents which have undergone an exchange. Females bearing the *nod* mutation and structurally normal *X* chromosomes produce approximately 3% *X* nondisjunction. Given that the frequency of nonexchange *X* tetrads in these experiments was approximately 6%, and that most (if not all) nondisjunctional events arose from nonexchange tetrads, Carpenter (1973) concluded that nonexchange *X* chromosomes disjoin at random in *nod* homozygous females.

This conclusion was recently confirmed when Ping Zhang, a student in my laboratory, induced fourteen new alleles of *nod* on a multiply inverted *X* chromosome called *FM7* (Zhang and Hawley, unpublished data). *FM7* is sufficiently rearranged so that in *FM7/+* heterozygotes the frequency of nonexchange *X* chromosomes approaches 100%. In *FM7*, *nod/nod* females, the frequency of *X* nondisjunction approaches 50% as would be predicted if nonexchange *X* chromosomes were disjoining at random.

Regardless of the level of *X* chromosome nondisjunction, females homozygous for *nod* also display a startlingly high level of fourth chromosome nondisjunction (85-90%). Most, if not all, of this apparent nondisjunction is due to chromosome loss. Distributively segregating chromosomes in nod females remain unstable well beyond meiosis I. Carpenter (1973) has observed, and we have confirmed, that nondisjunction of nonexchange X chromosomes can also occur at meiosis II. Nonexchange X and fourth chromosomes are also lost at high frequency during the early mitotic cleavage divisions. The loss of a maternal X chromosome during early zygotic cleavage divisions in an XX zygote leads to the production of XO/XX mosaics. Since XO tissue is male in Drosophila, such offspring develop as male/female mosaics known as gynandromorphs. The mitotic loss observed in the progeny of nod mothers is restricted to nonexchange chromosomes derived the nod mother, and thus may be considered an example of chromosomal imprinting.

The frequency with which a given chromosome, when nonexchange will undergo nod-induced mitotic loss seems to be related to chromosome size. Carpenter (1973) has noted that smaller chromosomes, such as the fourth chromosome, are lost at very high frequencies while large chromosomes such as the Y are rarely, if ever, lost. This relationship of instability and chromosome size has been confirmed in our own studies (Zhang and Hawley, unpublished data). For example, a small X chromosomal free duplication is lost in approximately 11% of the embryos which carry it, while a full length X chromosome is lost in only 3-5% of the embryos. The significance of these results remains unclear. Perhaps small chromosomes derived from nod mothers are less able to attach to the mitotic spindle. To elucidate this and other aspects of the nod phenotype, we have embarked on a molecular analysis of the nod locus.

The nod locus has been mapped precisely to band 10C2-3 on the standard polytene map of the

X chromosome. By a combination of genetic and molecular techniques, Ping Zhang has recently cloned the nod locus and mapped the RFLPs associated with a number of nod mutations. These mutations have allowed us to position the locus within 8kb of cloned DNA on the X chromosome of Drosophila (see Figure 2). The sequencing of this locus and the isolation of appropriate cDNAs is now underway. Our long term goals in this endeavor are to: 1) identify the nod^{+} product; 2) determine the timing of its synthesis, and perhaps most importantly, 3) to determine if the nod^{+} product is in fact a chromosomal protein.

CYTOLOGICAL STUDIES OF DISTRIBUTIVE PAIRING

Although the mechanism of distributive disjunction remains obscure, the cytological observations of Puro and Nokkala (1977) are suggestive. Early in metaphase I, achiasmate bivalents (such as the fourth chromosomes) co-orient on the same arc of the metaphase spindle and move precociously toward opposite poles. Indeed distributively segregating chromosomes were observed to reach the poles before chiasmate bivalents had fully separated. Similar phenomena have been observed cytologically in other insect meioses. For example, in Mantispid males, Hughes-Schrader (1969) documented a segregational mechanism in which the sex chromosomes and nonexchange autosomes (when present) segregate precociously along the same arc of the spindle without prior synapsis or physical connection; she has termed this phenomenon "distance segregation". A similar phenomenon has been documented in *Calocoris quandripunctatus* (Nokkala, 1986).

The mechanisms which underlie these disjunctions remain obscure. Perhaps the most instructive studies in this respect come from studies of achiasmate disjunction involving the sex chromosomes in the mole cricket *Gryllotalpa hexadactyla*. In this organism, two of the three

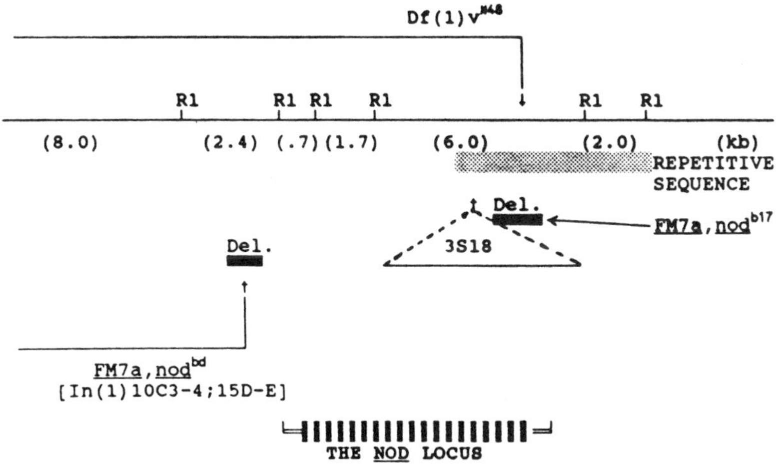

Figure 2: A restriction map of the nod locus. $Df(1)v^{N48}$ includes the nod locus and extends distally (towards the left of this diagram). FM7a, nod^{a} is an inversion with a breakpoint near to, but not within the nod locus. FM7a, nod^{b17} is a 0.5kb deletion. A cluster of spontaneously occuring nod mutations was recovered and shown to be the consequence of the insertion of a transposable element known as 3S18. A group of short repeated sequences is dispersed throughout this interval as indicated by the shaded region.

sex chromosomes normally segregate from each other faithfully without benefit of chiasmata. Indeed Camenzind and Nicklas (1968) have shown that passing a needle between these two chromosomes at metaphase I does not disrupt or prevent subsequent segregation. The mechanism of this segregation process may have been resolved by an ultrastructural analysis performed by Kubai and Wise (1981). Their work revealed that the centromeres of the X_1 and Y chromosomes were specifically modified such that a dark election-dense mass surrounded each of the two centromeres, and that centromere-to-centromere microtubules could be observed between these two chromosomes.

If, as seems likely, these centromere-to centromere microtubules ensure proper disjunction, then a simple mechanistic model for at least this type of achiasmate disjunction may be proposed: achiasmate chromosomes pair and then undergo centromere modification to allow their participation in an achiasmate system. Such modified centromeres are then capable of forming centromere-to-centromere tubules which ensure disjunction.

The strengths of applying such a model to distributive segregation in Drosophila are that it builds upon a well studied example, and that it explains the pleiotropic effects of a mutation, such as _nod_, on disjunction and chromosome stability. (One can easily imagine that _nod_ encodes a component necessary for proper centromere modification, and that without that component, centromere function is impaired at meiosis I, meiosis II, and early cleavage divisions.) The model, however, does not explain the mechanisms by which chromosomes choose disjunctional partners. A hypothesis to explain this process is proposed in a later section.

TOWARDS A MODEL OF DISTRIBUTIVE DISJUNCTION

Before proposing a model to account for distributive disjunction, two further issues must be considered. First, to what degree do those events which promote pairing for exchange overlap with those processes that ensure distributive segregations? Secondly, I will consider the degree to which distributive disjunctions involve pairs of chromosomes as opposed to the entire chromocenter or complex multivalents.

Is There Overlap Between The Distributive and Exchange Systems?

It is now almost axiomatic that exchange chromosomes do not normally enter into the distributive system. Indeed, the _ald_ mutation is specifically defective in this exclusion process. Moreover, the *nod* mutation, which basically randomizes the disjunction of distributively segregating chromosomes, has little or no effect on the disjunction of chromosomes which have undergone exchange (Carpenter, 1973). These studies have strongly argued that the mechanism(s) which ensures distributive disjunction is largely separate from the exchange system.

Nonetheless, several experiments suggest that the systems are not entirely independent. First, two lines of evidence suggest that *XX*<->*Y* or *XX*<->*44* pairing may decrease the frequency of exchange. Merriam (1967) showed that in Drosophila females homozygous for the *X* chromosomal inversion *In(1)w*^{m4}, the addition of a free *Y* chromosome suppresses recombination in the proximal region of the *X*. He argued that the exchange suppression results from competitive pairing between the two *X* and *Y* chromosomes near the base of the *X* (i.e. near the site of *XXY* pairing). Similarly, in females bearing the *Axs* mutation, which suppresses homolog segregation in the distributive system and promotes heterologous disjunction (see above), Anne Zitron has observed a weak but consistent decrease in proximal *X* chromosomal exchange. No such effects were seen more distally. It is possible that this effect is analogous to that observed by Merriam in *XXY* females, and thus reflects an exchange suppression due to the heterologous (*X*-*4*) associations promoted by the *Axs* mutation.

The second group of experiments demonstrates that exchange chromosomes can participate in distributive disjunctions. For example, Carpenter (1973) has shown that in *XXY* females exchange chromosomes do occasionally participate in *XX*<->*Y* segregations (i.e. secondary nondisjunction). Indeed, Carpenter found that fully 25% of the *XX*<->*Y* segregations observed in her crosses involved exchange bivalents. Although lower frequencies of *XX*<->*Y* disjunctions involving exchange chromosomes

(3%) were observed by Bridges (1916) and others, it is nonetheless clear that exchange X chromosomes are not entirely prevented from engaging in XX<->Y distributive disjunctions. Similarly, Novitski (1978) has shown that the Y chromosomal arms of a Y-autosome translocation (T(Y;A)) can still orient the disjunction of two nonexchange X chromosome, even when the autosomal material connected to the Y is involved in exchange.

Finally, the mei-S51 mutation both disrupts the exchange process and interferes with proper distributive disjunction, suggesting a common component to the systems.

These data suggest that there is at least some degree of overlap between those events that ensure proper homologous pairing and exchange, and those chromosome alignments which will (in the absence of chiasmata) result in distributive disjunction. Moreover, as reviewed by Novitski and Puro (1978), detailed cytological studies on Drosophila have failed to provide convincing evidence of a second round of pairing as proposed by Grell (1962a, 1962b, and 1976).

Rather, the majority of data appear to support Novitski (1964, 1975, and 1978) and indicate that the chromosomal associations which underlie both homologous pairing for exchange and pairing for distributive disjunction are established early during meiosis. Novitski (1964) further proposed that most, if not all, of the associations leading to distributive segregation involve sequences near the centromeres. For normal sequence chromosomes, such regions would be heterochromatic, and thus unable to participate in exchange events. (One basis for such associations might be the homology between repetitive sequences, such as satellite DNA and moderately repetitive sequences, which are distributed throughout the heterochromatic arms.)

In other words, the same mechanism(s) which

optimize meiotic pairing within the euchromatin may act to facilitate chromosome alignment within the chromocenter. We postulate that the specificity of such chromocentral pairings or alignments is regulated by genes such as mei-S51 and Axs, and possibly ald. Indeed as we have noted above, mei-S51 may define a common component of both nonexchange alignment and euchromatic homologous pairing.

The Role of Multiple Chromosome Associations (Multivalents) in Distributive Disjunctions.

The primary segregational event within the distributive system appears to be the disjunction of either homologous or heterologous pairs of chromosomes. This is exemplified by the cytological analysis of the c(3)G mutation which completely eliminates exchange.

In females homozygous for c(3)G the chromosomes tend to associate in pairs during early prometaphase. However, by mid-prometaphase the bivalents separate prematurely, such that all chromosomes appear as univalents (Puro and Nokkala, 1977). At anaphase I, these univalents proceed precociously to the poles in a fashion analogous to that seen for normal distributive disjunctions (i.e. for the fourth chromosomes). Segregation is variable, although the existing cytological data suggest that approximately equal numbers of chromosomes proceed to each pole in variable combinations. These observations are consistent with the genetic analysis of c(3)G by Hall (1972) who observed a substantial tendency of nonhomologous chromosomes to disjoin from each other in c(3)G females. This was especially true for the larger chromosomes. Thus, most of the nondisjunction observed can be accounted for by heterologous disjunctions.

Although these data suggest that bivalent formation is the primary means of ensuring disjunction, we have noted above that XX<->Y distributive disjunction in XXY females is mediated by trivalent formation. As mentioned earlier, there is also an extensive literature on other nonhomologous disjunctions within the distributive system which demonstrate the ability of a nonexchange metacentric to orient two nonexchange acrocentric chromosomes (for review see Grell, 1976). These observations may suggest that distributive associations need not be restricted to pairs of chromosomes but rather, that nonexchange associations act to minimize the extent of unpaired chromatin within the chromocenter.

Moreover, the analysis of distributive disjunction in the presence of several mutations strongly suggests that the formation of complex multivalents, and their subsequent segregation, represents a default process within the distributive system. For example, in Axs or ald XXY females the most reasonable interpretation of the observed segregation is the formation of an XXY44 multivalent involving nonexchange X chromosomes. Thus, within the chromocenter each chromosome (or heterochromatic interval) may be involved in multiple associations with both its homolog and with other nonhomologous chromosomes.

One can imagine then that two parallel systems exist for resolving this chromosomal bouquet into separate bivalents (or in some cases trivalents). One system is chiasmate in nature and relies on the ability of chiasmata to link together homologous centromeres. The other, distributive segregation, resolves the nonexchange chromosomes into bivalents, or trivalents, on the basis of (1) availability, (2) homolog recognition (if possible), (3) size, and (4) shape.

A NEW MODEL OF DISTRIBUTIVE DISJUNCTION

Following Novitski (1964, 1975, and 1978), Anne Zitron and I have proposed that distributive associations occur concommitantly with pairing for euchromatic exchange and are primarily based on homology. To quote Cooper (1945) "...the primary mechanism which brings about synapsis itself may likewise provide a device ensuring segregation if the paired condition is maintained until metaphase. Very possibly the paired synaptic state can persist to metaphase in organisms having very small or little-coiled chromosomes at meiosis whether or not chiasmata are formed". We would further argue that because of the multiple homologies between heterochromatic regions, such pairing can also involve nonhomologous chromosomes, although with less efficacy.

The proposed dependence of the distributive system on homology raises two important questions. The first question is, why do two nonexchange *X* chromosomes and a *Y* chromosome routinely form trivalents?. The answer to this question may lie in the limited homology between the *X* and *Y* chromosomes. This homology results from the presence of common satellite DNA sequences and ribosomal DNA on both *X* and *Y* chromosomes.

The second, and perhaps more vexing question, relates to the interactions between the *X* and fourth chromosomes. I have noted above that *X* duplications of appropriate size routinely interfere in fourth chromosomal disjunctions, and that *XX*<->*44* disjunctions are a common phenotype of mutations that affect distributive disjunction. Moreover, although the numbers are low, Drosophila females which are heterozygous for *X* chromosomal inversions exhibit a trend towards *XX*<->*44* disjunction when both the *X* and fourth chromosomes are nondisjunctional. The explanation of these anomalies may be found in a considerable amount of genetic and cytological evidence suggesting that the *X* and fourth chromosomes are evolutionarily related (for review see Hochman, 1976). In addition, Miklos *et al*. (1988) have recently shown that the fourth chromosome and the proximal euchromatin of the _X_ are highly homologous with respect to the number, composition, and distribution of certain classes of repetitive sequences. Thus, homologous sequences present on both the _X_ and fourth chromosomes may provide frequent opportunities for _X-4_ distributive associations than can result in improper segregation events.

If this portion of our model is correct, the term "distributive pairing" corresponds in reality to chromocentral pairings involving both homologous and nonhomologous chromosomes. We propose that the cell attempts to maximize the fidelity of such associations by pairing those chromosomal regions with the greatest homology. These associations help to ensure that each chromosomal arm will be paired (perhaps by utilizing the limited homology common to heterochromatic intervals). The latter assumption allows us to understand both _XXY_ pairings (based on limited homology) and associations between less homologous chromosomes. (In this sense the control of distributive chromosomal associations may be analogous to the mechanisms which regulate homologous versus homeologous pairing in wheat; for review see Sears, 1976). As suggested above, such pairings may indeed be sufficient to ensure disjunction in the absence of chiasmata.

As aesthetically pleasing as such a model may be, at least one major aspect of distributive disjunction remains unexplained; namely, there is no obvious reason that chiasmata should preclude chromosomes from participating in the distributive system. Why should chiasmata override alignments formed in the chromocenter?

The solution to this dilemma is the observed precocious disjunction of nonexchange bivalents (see Nokkala and Puro, 1978). We propose that distributive segregations represent the first step in meiotic segregation, and that exchange chromosomes fail to participate in such disjunctions precisely because chiasmata hold them at the metaphase plate. As heterodox as such a proposal may seem, it is in good accord with the available cytological evidence.

Our model differs from those of Grell (1962a) and Novitski (1974) in that it proposes that distributive segregations occur prior to the disjunction of exchange chromosomes. If this model is correct then the formal model of meiosis presented above must be restated as follows. Both euchromatic pairing and chromocentral alignment occur during the early stages of meiotic prophase followed by exchange in the euchromatic regions. Although euchromatic and chromocentral pairings both occur during early meiosis, they are usually not competitive because most of the chromocentral material is heterochromatic in nature. At the beginning of metaphase I, nonexchange chromosomes are free to disjoin from those chromosomes with which they are most intimately associated. The mechanisms of such disjunction remain undefined, but the phenotypes of <u>nod</u> suggest that some sort of chromosomal or centromeric modification may well be required for distributive disjunction. Chiasmate bivalents will be held together on the metaphase plate, and thus prevented from disjoining at this time. The final step of the model is the actual disjunction of exchange chromosomes and anaphase I.

According to this model, all chromosomes would participate in the chromosomal associations and alignments that define the patterns of distributive segregation. Those associations, such as <u>X</u>-<u>Y</u>-<u>X</u> pairings, involving exchange bivalents would normally not induce nondisjunction because the chiasma would prevent the exchange chromosomes from leaving the metaphase plate until anaphase. Distributive disjunctions involving nonexchange chromosomes would, however, allow precocious disjunction at metaphase I. Because such disjunctions are based upon homology dependent pairings, most distributive disjunctions would involve homologs. Normally heterologous disjunctions, those inducing nondisjunction, would occur only when proper homolog pairing was prevented (i.e. in the absence of a homolog) or when the presence of an unpaired chromosome, such as a *Y*, has resulted in improper pairings.

This model explains the interactions between the distributive and exchange systems described above. For example, given that exchange pairings and distributive associations are proposed to occur at the same time, one can easily see how the *Y* chromosomal arms of a *T(Y;A)* might orient nonexchange *X* chromosomes (as reported by Novitski 1975, but see above), even when the autosomal material has undergone exchange. Similarly, the suppressive effects of the *Y* chromosome on recombination at the base of the *X* can also be understood by competitive pairing (i.e. *XX*<->*Y* vs *X*<->*X*).

Although this modification of the Novitski (1964) model is speculative, it does make some useful predictions. For example, one might predict that, in addition to *mei-S51*, other meiotic mutants that affect the establishment of pairing, or other preconditions for exchange, might also affect distributive disjunctions. Experiments to test this and other predictions of this model are underway in our laboratory.

ACKNOWLEDGEMENTS

The author acknowledges the support of a grant from the Eukaryotic Genetics Program of the National Science Foundation (DCB-8815749). The author is also the recipient of a Faculty Research Award from the American Cancer Society (FRA-324) and of an Irma T. Hirschl Monique-Weill Caulier Career Scientist Award. He also acknowledges the kindness of Anne Zitron and Ping Zhang for sharing their unpublished data, and Anne Zitron, Brenda Brodeur, and Ping Zhang for valuable discussion. Finally, the author is greatly indebted to Donna Lombardi for her extraordinary patience and diligent efforts in the preparation of this manuscript.

REFERENCES

Baker, B.S., and J.C. Hall, 1976. Meiotic mutants: genetic control of meiotic recombination and chromosome segregation. In Genetics and Biology of Drosophila Vol. 1A, edited by M.A. Ashburner and E. Novitski; Academic Press, New York and London, pp. 351-434.

Bridges, C.B., 1916. Nondisjunction as the proof of the chromosome theory of heredity. Genetics 1: 1-51 and 107-163.

Carpenter, A.T.C., 1973. A mutant defective in distributive disjunction in Drosophila melanogaster. Genetics 73: 393-428.

Camenzind and R.B. Nicklas, 1968. The non-random chromosome segregation in spermatocytes of Gryllotalpa hexadactyla. A micromanipulation analysis. Chromosoma (Berl.) 24: 324-335.

Cooper, K.W., 1945. Normal segregation without chiasmata in female Drosophila melanogaster. Genetics 30: 472-484.

Cooper, K.W., 1948. A new theory of secondary nondisjunction in female Drosophila melanogaster. Proc. Natl. Acad. Sci. U.S.A. 34: 179-187.

Gershenson, S., 1940. The nature of the so-called genetically inert parts of chromosomes. Videnskap. Akad. Nank. U.R.R.S. 3: 116.

Grell, R.F., 1962a. A new hypothesis on the nature and sequence of meiotic events in the female of Drosophila melanogaster. Proc. Natl. Acad. Sci. U.S.A. 48: 165-172.

Grell, R.F., 1962b. A new model for secondary nondisjunction: the role of distributive pairing. Genetics 50: 150-166.

Grell, R.F., 1976. Distributive pairing. In Genetics and Biology of Drosophila melanogaster Vol 1A, edited by M.A. Ashburner and E. Novitski; Academic Press, New York and London, pp. 435-486.

Hall, J.C., 1972. Chromosome segregation influenced by two alleles of the meiotic mutant c(3)G in Drosophila melanogaster. Genetics 71: 367-400.

Hawley, R.S., 1988. Exchange and chromosomal segregation in eukaryotes. pp. 497-525. In: Genetic Recombination, Edited R. Kucherlapati and G. Smith. American Society for Microbiology, Washington, D.C.

Hochman, B., 1976. The fourth chromosome of Drosophila melanogaster. In: Genetics and Biology of Drosophila Vol. 1B, edited by M.A. Ashburner and E. Novitski. Academic Press, New York and London, pp. 903-928.

Hughes-Schrader, S., 1969. Distance segregation and compound sex chromosomes in mantispids. Chromosoma 27: 109-129.

Kubai, D.F. and D. Wise, 1981. Nonrandom chromosome segregation in Neocurtilla (Gryllotalpa) hexadactyla. J. Cell Biology 88: 281-293.

Lindsley, D.L. and L. Sandler, 1958. The meiotic behavior of grossly deleted X chromosomes in Drosophila melanogaster. Genetics 43: 547-563.

Merriam, J.R., 1967. The initiation of nonhomologous chromosome pairing before exchange in female Drosophila melanogaster. Genetics 57: 409-425.

Miklos, G.L.G., M.T. Yamamoto, J. Davies and V. Pirrotta, 1988. Microcloning reveals a high frequency of repetitive sequences characteristic of chromosome 4 and the beta-heterochromatin of Drosophila melanogaster. Proc. Natl. Acad. Sci. U.S.A. 85: 2051-2055.

Nokkala, S., 1986a. The mechanisms behind the regular segregation of autosomal univalents in Calocoris quandripunctatus. Hereditas 105: 199-204.

Novitski, E., 1964. An alternative to the distributive pairing hypothesis in Drosophila. Genetics 50: 1449-1451.

Novitski, E., 1975. Evidence for the single phase pairing theory of meiosis. Genetics 79: 63-71.

Novitski, E., 1978. The relation of exchange to nondisjunction in heterologous chromosome pairs in the Drosophila female. Genetics 88: 499-503.

Novitski, E. and J. Puro, 1978. A critique of theories of meiosis in the female of Drosophila melanogaster. Hereditas 89: 51-67.

O'Tousa, J., 1982. Meiotic chromosome behavior influenced by mutation-altered disjunction in Drosophila melanogaster females. Genetics 102: 503-524.

Puro, J. and S. Nokkala, 1977. Meiotic segregation of chromosomes in Drosophila melanogaster oocytes. Chromosoma 63: 273-286.

Robbins, L.G., 1971. Nonexchange alignment: a meiotic process revealed by a synthetic meiotic mutant of Drosophila melanogaster, Mol. Gen. Genet. 110: 144-166.

Sears, E.R., 1976. Genetic control of chromosome pairing in wheat. Ann. Rev. Genet. 10: 31-53.

Weinstein, A., 1936. The theory of multiple-strand crossing over. Genetics 21: 155-199.

Molecular and Cytogenetic Studies of
Non-Disjunction, pages 303–326

ISOLATION AND CHARACTERIZATION OF YEAST MUTANTS DEFECTIVE IN MEIOTIC CHROMOSOME SEGREGATION

G. Shirleen Roeder, Beth M. Rockmill, JoAnne Engebrecht, Emily A. Thompson and Thomas M. Menees

Department of Biology, Yale University, New Haven, Connecticut 06511-8112

INTRODUCTION

In *Saccharomyces cerevisiae*, several genes have been demonstrated to be essential for meiotic, but not mitotic, chromosome transmission. These genes are defined by mutants defective at various steps in meiosis including the pairing of homologous chromosomes, initiation of genetic exchange, resolution of recombination intermediates and meiosis I disjunction.

Mutations at the *SPO12* and *SPO13* loci cause cells to skip the meiosis I reductional division and undergo a single round of equational chromosome segregation (Klapholz and Esposito, 1980). Consequently, these mutants produce dyads containing diploid spores instead of tetrads containing haploids. The *spo12* and *spo13* mutants undergo normal chromosome pairing and promote wild-type levels of meiotic recombination. A *spo13* mutation can correct the spore lethality associated with some recombination-defective meiotic mutants suggesting that gene products normally required for proper meiosis I disjunction are dispensable in a *spo13* background (Malone and Esposito, 1981; Malone, 1983). A *spo13* mutation also permits haploid cells to undergo meiosis, provided these cells express both *a* and *α* mating type information (Wagstaff et al., 1982).

A mutant defective in the pairing of homologous chromosomes has recently been described by Hollingsworth and Byers (1989). Mutations at the *HOP1* locus reduce meiotic interchromosomal recombination approximately ten-fold but have no affect on intrachromosomal exchange. Electron-microscopic observations indicate that *hop1* mutants fail to assemble synaptonemal complex (SC). Thus, *hop1* mutants appear to be defective in the pairing of homologous chromosomes, a necessary prerequisite to interhomologue, but not intrachromosomal, exchange. *hop1* mutants produce inviable spores but this lethality is rescued by a *spo13* mutation; i.e., *hop1 spo13* strains produce viable progeny.

Mutations at the *RAD50* and *SPO11* loci result in reduced sporulation efficiencies and the production of inviable spores (Game et al., 1980; Klapholz et al., 1985). The spores produced by *rad50 spo13* and *spo11 spo13* double mutants are viable but nonrecombinant indicating that the *RAD50* and *SPO11* gene products are required for meiotic exchange (Malone and Esposito, 1981). SC fails to assemble in *rad50* mutants (Farnet et al., 1989) and a *spo11* deletion mutant (Giroux, 1989); however, there is some SC in the *spo11-1* (presumed point) mutant (Klapholz et al., 1985). It is possible that the *RAD50* and *SPO11* genes encode enzymes involved in meiotic recombination and that the SC is unstable in the absence of these proteins. Alternatively, one or both of these genes may encode structural components of the SC. Finally, it is possible that the *RAD50* and/or *SPO11* gene products are regulatory proteins that control the expression or activity of other genes or gene products required for SC formation and meiotic recombination.

The *rad51*, *rad52*, *rad55* and *rad57* mutants also produce inviable spores (Game et al., 1980) but these cannot be rescued by a *spo13* mutation. The recombination competence of the *rad52* and *rad57* mutants has been examined in "return-to-growth" experiments in which cells are induced to enter meiosis and undergo meiotic recombination and then returned to growth medium where they resume vegetative division. Under these conditions, wild-type cells display meiotic levels of recombination but do not undergo meiotic chromosome segregation (Esposito and Esposito, 1974). In the return-to-growth protocol, the *rad52* and *rad57* mutants display reduced rates of meiotic exchange (Game et al., 1980; Resnick et al., 1986). However, analysis of these mutants by a physical recombination assay indicates that both produce significant amounts of recombinant restriction fragments (Borts et al., 1986). It is generally assumed that the *rad52* and *rad57* mutants (and probably also *rad51* and *rad55*) initiate recombination normally but are defective in the resolution of recombination intermediates. Consistent with this interpretation, *spo11* or *rad50* mutations (that confer a failure to initiate exchange) can rescue the lethality of *rad52 spo13* and *rad57 spo13*; i.e., *spo11* (or *rad50*) *rad52* (or *rad57*) *spo13* triple mutants produce viable nonrecombinant progeny (Malone, 1983; R. E. Esposito, personal communication).

The genes just described probably represent a small fraction of the total required for the proper transmission of chromosomes during meiosis. We have carried out screens for new meiosis-defective mutants in an attempt to identify additional genes involved in meiotic chromosome segregation.

RESULTS AND DISCUSSION

A Screen for Meiotic Lethal Mutants

Yeast mutants defective in meiotic chromosome segregation can sporulate but the meiotic progeny are inviable due aneuploidy. Thus, a search for sporulation-proficient, meiotic-lethal mutants was conducted. In order to recover recessive mutations in a diploid strain, haploid spores from a homothallic (*HO*) diploid were mutagenized (Esposito and Esposito, 1969). Homothallic spores switch mating type within the first few cell divisions and then cells of opposite mating type, coexisting within a colony, mate with each other to generate diploids homozygous for any induced mutations (for review, see Nasmyth, 1982). The haploid meiotic products from an *HO* diploid strain were plated on solid medium and exposed to ultraviolet (UV) light at a dose sufficient to achieve 50 % killing. The surviving spores gave rise to diploid colonies which were then replica plated to sporulation medium to induce meiosis.

Because yeast does not sporulate with 100 % efficiency, spore viability cannot be assessed simply by examining the growth of a sporulated colony. The growth of unsporulated diploids obscures any reduction in spore viability. This problem was overcome by taking advantage of the differential sensitivity to ether vapors of vegetative cells compared to spores (Dawes and Hardie, 1974). Sporulated colonies were replica plated to complete medium and then briefly exposed to ether vapors. Colonies which did not sporulate or produced mostly inviable spores failed to grow after the exposure to ether. Prospective mutants were examined in the light microscope to assess sporulation proficiency. After screening approximately 30,000 mutagenized cells, thirteen sporulation-competent, meiotic-lethal mutants were recovered.

To construct mating-competent haploid strains carrying the meiotic-lethal mutations, spores from each of the mutants were mated to *ho* haploids. (Even though most of the spores are dead, a few percent are viable and mating-competent.) The resulting diploids, heterozygous for *ho* and a meiotic-lethal mutation, were sporulated and tetrads were dissected and analyzed. In this way, *ho* mutant haploids were isolated and these were used for complementation testing. From the thirteen mutants, two pairs of noncomplementing alleles were recovered; thus, eleven different complementation groups are represented by the thirteen mutants. All of the mutants complement the previously described meiotic-lethal mutants.

Two of the mutations recovered represent alleles of the *RED1* gene (for reductional division), which maps to chromosome XII. Two other mutations define the *MER1* gene(for meiotic recombination), which is tightly linked to the *PETX* locus on chromosome XIV. The following sections present the results of genetic studies of the *red1-1* and *mer1-1* mutants and molecular studies of the cloned *RED1* and *MER1* genes. Genetic studies of the *red1-1* mutant have previously been presented by Rockmill and Roeder (1988); the expression and DNA sequence of the *RED1* gene has been described by Thompson and Roeder (submitted). Characterization of the *MER1* gene has been reported by Engebrecht and Roeder (1989).

red1-1 Spores are Highly Aneuploid

The *red1-1* mutant sporulates efficiently but only 1 % of the spores produced are viable. If the spore inviability observed is the consequence of chromosome nondisjunction, then the rare viable spores are expected to be highly aneuploid. To test this hypothesis, the meiotic products of a *red1-1* diploid were examined for disomy of chromosomes III and VIII. The diploid strains used for this analysis carried codominant alleles at the *MAT* locus (*MATa/MATα*) on chromosome III and the *ARG4* locus (*arg4-8/arg4-9*) on chromosome VIII. Spores that inherit a single copy of chromosome III are either *a*- or *α*-maters; aneuploid spores that carry both *MATa* and *MATα* information are nonmaters. The *arg4-8* and *arg4-9* alleles display intragenic complementation. Haploid spores that inherit either one of the two alleles are arginine auxotrophs whereas those aneuploid spores that carry both alleles are prototrophs.

Among the spores produced by *RED1* diploids, no nonmaters were detected and only 1-2 % Arg^+ spores were recovered. The latter were haploids and were Arg^+ as a result of intragenic recombination. Of the spores produced by a *red1-1* homozygous diploid, 34 % were nonmaters and 33 % were Arg^+.

These results demonstrate that there are high levels of aneuploidy among the survivors of meiosis in the *red1-1* mutant and indicate that the spore inviability observed is the consequence of meiotic nondisjunction.

Nondisjunction in the *red1-1* Mutant Occurs Exclusively at the First Meiotic Division

To determine whether the nondisjunction that occurs in a *red1-1* mutant takes place at the first and/or second meiotic divisions, spores disomic for chromosome VIII were selected and analyzed for the presence of sister or nonsister chromosomes. Spores disomic for chromosome VIII can be selected on the basis of resistance to copper which is mediated by the *CUP1* gene. The degree of copper-resistance is proportional to the number of copies of *CUP1;* thus, spores disomic for chromosome VIII (1n + 1) can be distinguished from haploids (1n) because they are resistant to higher concentrations of copper.

Nondisjunction at meiosis I results in spores carrying one copy of each of the two parental chromosomes whereas nondisjunction at meiosis II generates disomic spores carrying two copies of the same chromosome (Figure 1). These possibilities were distinguished by marking the parental chromosomes with *arg4-8* and *arg4-9* heteroalleles. Spores that inherit one copy of each chromosome are Arg^+ but those that inherit two copies of the same chromosome are Arg^- (Figure 1). The results of this analysis indicate that most or all of the aberrant segregation that occurs in the *red1-1* mutant occurs at the first meiotic division (for details, see Rockmill and Roeder, 1988).

red1-1 Spore Inviability is Rescued by *spo13*

As discussed in the Introduction, a *spo13* mutation can rescue the spore inviability associated with some mutants defective in meiosis I. To determine whether a *spo13* mutation can

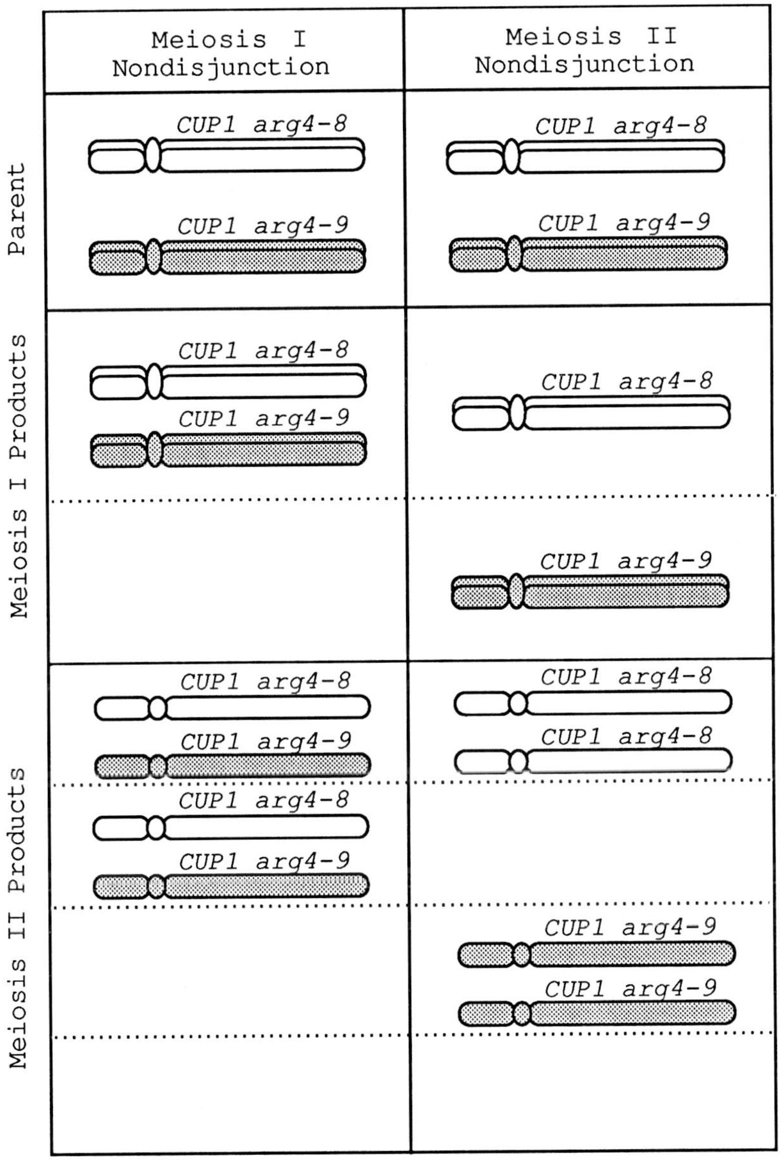

Figure 1. Nondisjunction of chromosome VIII at meiosis I or II. Meiosis I nondisjunctants are copper-resistant and prototrophic; meiosis II nondisjunctants are copper-resistant auxotrophs.

rescue *red1* mutants, *red1-1 spo13* diploids were constructed and analyzed for spore viability by dyad dissection. Both *RED1 spo13* and *red1-1 spo13* homozygous diploids produced 82 % viable spores, indicating that a *spo13* mutation can restore viability to the *red1-1* mutant. This result provides further evidence that the *red1-1* mutant is defective at the first meiotic division.

The *red1-1* Mutant is Recombination-Proficient

Genetic recombination is a necessary prerequisite to the proper segregation of chromosomes at the first meiotic division. Thus, the nondisjunction observed in the *red1-1* mutant might result from a defect in meiotic exchange. To assay recombination in the *red1-1* mutant, dyads from *red1-1 spo13* diploids were dissected and examined for crossing over in the *LEU2-MAT* and *HIS4-LEU2* intervals on chromosome III. As shown in Figure 2, map distances in wild-type and mutant are similar, indicating that intergenic crossing over occurs at normal levels in the *red1-1* mutant.

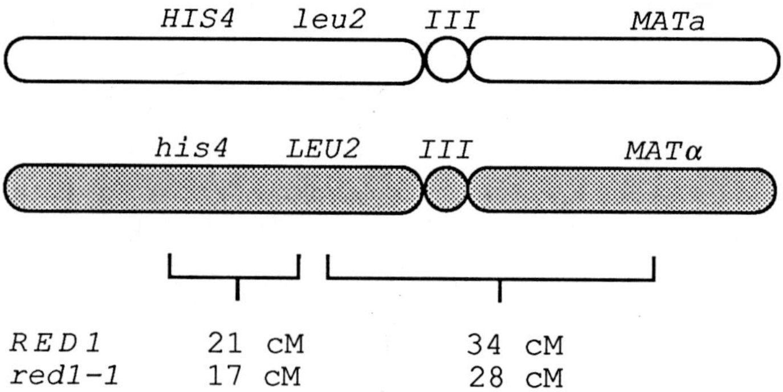

Figure 2. Map distances in wild type and *red1-1*.

Cloning and Sequencing of the *RED1* Gene

The wild-type *RED1* gene was isolated from a yeast genomic library based on complementation of the spore inviability phenotype. To locate the *RED1* gene on the cloned DNA segment, transposon insertion mutations were generated in *E. coli* and then transformed into yeast according to the shuttle mutagenesis scheme described by Seifert et al. (1986). This analysis localized *RED1* to an interval of approximately 3.0 kilobases (kb) (Figure 3).

In an attempt to gain information about the structure and function of the *RED1* protein, the DNA sequence of the gene was determined. This analysis identified a 2484 bp open reading frame whose position corresponds to the *RED1* region defined by transposon mutagenesis. The predicted gene product is 827 amino acids in length with a molecular weight of 95.5 kilodaltons. The *RED1* protein is highly charged with 11.9 % lysine, 9.4 % glutamine, 6.0 % aspartic acid, 3.4 % arginine and 1.8 % histidine residues for a total of 32.5 % charged amino acids.

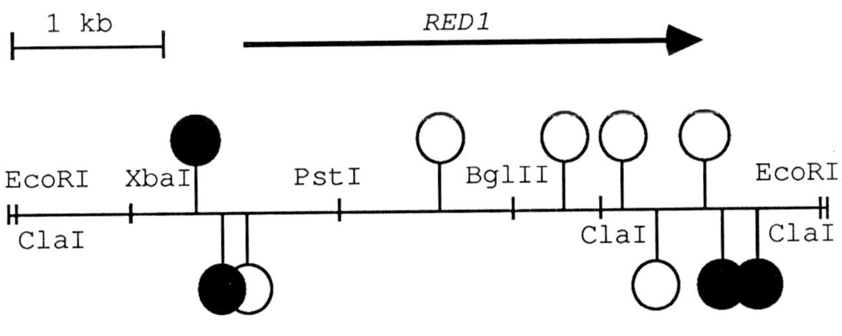

Figure 3. Restriction map of a 5.2 kb EcoRI fragment containing the *RED1* gene. Filled circles indicate transposon insertions that confer a Red$^+$ phenotype; open circles indicate Red$^-$ mutants. The arrow above the line indicates the position and orientation of the major *RED1* transcript.

The *RED1* amino acid sequence was compared with others in the SWISS-PROT databank and the Protein Identification Resource. No significant homologies were found. In addition, comparisons were made between the *RED1* protein and the products of nine genes known to be essential for meiosis and/or expressed specifically in meiotic cells. These include *SPO11* (Atcheson et al., 1987), *SPO13* (R. E. Esposito, personal communication), *RAD50* (E. Alani, S. Subbiah and N. Kleckner, personal communication), *RAD52* (Adzuma et al., 1984), *HOP1* (N. Hollingsworth and B. Byers, personal communication), *CDC31* (Baum et al., 1986), $mei3^+$ (McLeod et al., 1987) *SPS4* (Garber and Segall, 1986), *SPS2* and a fragment of *SPS1* (Percival-Smith and Segall, 1986). All of these genes are from *S. cerevisiae* except for $mei3^+$ which is from *Schizosaccharomyces pombe*. In no case were any significant homologies detected.

Expression of the *RED1* Gene is Meiotically-Induced

To study transcription of the *RED1* gene, Northern blot hybridization was used to analyze RNA from mitotic cells and cells harvested at various timepoints after transfer to sporulation medium. Two RNAs were detected with the *RED1* probe, a major transcript of 2.75 kb and a minor species of 3.2 kb. The major 2.75 kb transcript was not detected in mitotically-dividing cells. It was first apparent at 2.5 hr after transfer to starvation conditions, accumulated maximally between 5 and 7.5 hr and then decreased. The minor transcript accumulated in meiosis with a profile similar to that of the major transcript. However, the 3.2 kb RNA is only about 5 % as abundant as the 2.75 kb transcript.

The Northern blot shown in Figure 4 was hybridized with both *RED1* and *SPO13* probes to allow comparison of the timing of *RED1* and *SPO13* gene expression. The major 1.0 kb *SPO13* transcript (Wang et al., 1987) accumulates with kinetics similar to those of *RED1* suggesting that these genes are coordinately regulated.

To correlate the timing of *RED1* transcript accumulation with other events in meiosis, cells from each timepoint were examined for chromosome segregation by DAPI staining. Binucleate cells, products of the first meiotic division, reached maximum abundance at 10 hours, just a few hours after the peak of *RED1* transcript accumulation. Thus, the timing of *RED1* expression corresponds approximately with the time in meiosis at which the *RED1* defect is apparent.

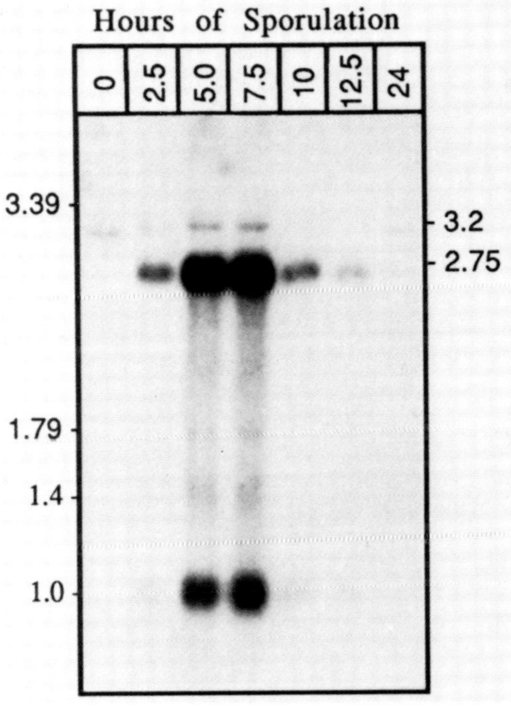

Figure 4. Transcription of *RED1* and *SPO13*. Shown is a Northern blot of total RNA prepared from strain SK1 at the indicated times after transfer to sporulation medium. The blot was hybridized with antisense RNA probes for both *RED1* and *SPO13*. The positions and sizes of the 25S (3.39) and 18S (1.7) rRNAs and the *RED1* (3.2 and 2.75) and *SPO13* (1.4 and 1.0) transcripts are indicated.

The Role of the *RED1* Gene Product

Several possibilities for the function of the *RED1* protein are consistent with the *red1-1* mutant phenotype. First, the *red1-1* mutant could be defective in kinetochore function. Chromosomes might fail to attach to the meiosis I spindle or kinetochores might differentiate prematurely such that sister chromatids separate from each other prior to the meiosis I division. Second, the *RED1* protein might be required for structure or function of the meiosis I spindle pole body or spindle. Third, *RED1* might play a role in chiasma function. After the SC dissolves, homologous chromosomes remain attached to each other at chiasmata, which correspond to the sites of genetic crossovers. The *red1-1* mutant might be defective in the maintenance of chiasmata such that homologues separate from each prematurely. Alternatively, the *red1-1* mutant might be defective in the resolution of chiasmata such that homologues are unable to disjoin from each other.

Two experimental approaches to determining the function of the *RED1* gene product are in progress. First, second-site suppressor mutations of *red1* mutants are being isolated in an attempt to identify genes whose products interact with the *RED1* protein. Second, antibodies to the *RED1* gene product are being used to localize the *RED1* protein within meiotic cells by immunolabeling. Localization of the *RED1* protein to a particular subcellular structure such as the SC, spindle, kinetochore, spindle pole body or chiasma will provide significant insight into the mechanism of *RED1* action.

Characterization of the *mer1-1* Mutant

The *mer1-1* mutant sporulates efficiently but only 2 % of the spores produced are viable. *mer1-1 spo13* double mutants produce viable products indicating that the *mer1-1* spore lethality can be rescued by a *spo13* mutation. Crossing over in the *HIS4-MAT* interval on chromosome III was measured in diploids homozygous for the *mer1-1* and *spo13* mutations. The *HIS4-MAT* distance measured 42 cM in the wild-type control but only 2.9 cM in the *mer1-1* mutant. Thus, recombination is reduced, but not completely eliminated, by the *mer1-1* mutation.

Cloning, Disruption and Expression of *MER1*

The partial recombination defect observed in the *mer1-1* mutant suggested the possibility that *mer1-1* may be a leaky mutation and that null mutations at the *MER1* locus might completely eliminate meiotic exchange. To address this possibility, the *MER1* gene was cloned and strains carrying deletion mutations were constructed.

The wild-type *MER1* gene was isolated from a yeast genomic library by complementation of the meiotic-lethal phenotype. The gene was localized to a region of approximately 1.0 kb by subcloning and transposon mutagenesis (Figure 5). Two deletion/disruption mutations were constructed and used for subsequent studies of the *mer1* mutant phenotype. In one of these, a fragment containing the *LYS2* gene was used to replace approximately 300 bp of *MER1* sequence. In the other, 600 bp of *MER1* sequence were deleted and replaced by *LEU2* (Figure 5).

The transposon used for insertional mutagenesis of *MER1* carries all of the *E. coli lacZ* gene, except the first eight amino acids (Seifert et al., 1986). The *lacZ* coding region is fused in-frame to an open reading frame in one of the terminal repeats of the transposon; thus, an open reading frame extends from one end of the element through *lacZ*. When this transposon inserts into *MER1* in the correct orientation and proper reading frame, a fusion gene producing a *MER1-lacZ* chimeric protein results. Assays of β-galactosidase activity in yeast strains carrying *MER1-lacZ* fusions indicate that expression of the *MER1* gene is meiotically-induced with kinetics similar to those observed for *RED1* and *SPO13*.

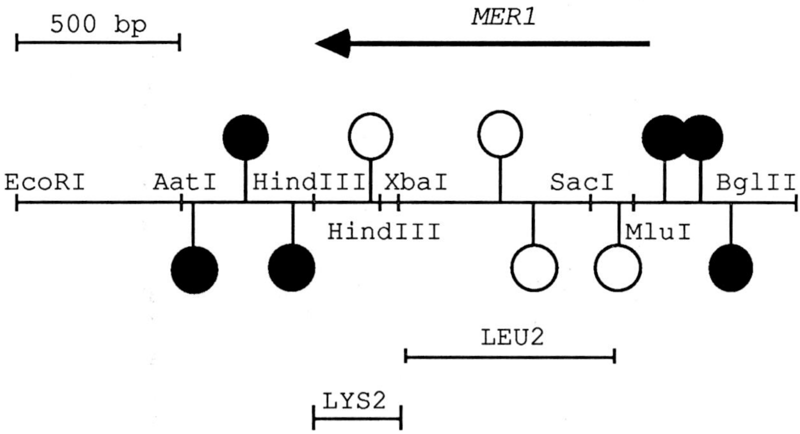

Figure 5. Restriction map of a 2.4 kb EcoRI-BglII fragment containing the *MER1* gene. Filled circles indicate insertion mutations that confer a Mer^+ phenotype; open circles indicate Mer^- mutants. The arrow indicates the direction of *MER1* transcription as demonstrated by *MER1-lacZ* fusion genes. The endpoints of the *LYS2* and *LEU2* deletion/disruption mutations are indicated.

mer1 Mutants Are Defective in Both Intragenic and Intergenic Recombination between Homologues

Isogenic *spo13* strains with and without the *mer1::LEU2* null allele were used to assay crossing over in the *HIS4-MAT* interval on chromosome III and the *CYH10-LYS2* region on chromosome II. Crossing over was decreased to approximately 10 % of wild-type levels in the *mer1* deletion mutant (Figure 6). This reduction is similar to that observed for the *mer1-1* mutation, suggesting that the UV-induced allele is a null mutation.

The effect of a *mer1* null mutation on nonreciprocal gene conversion events was examined in *spo13* strains carrying *his4* and *leu2* heteroalleles. His$^+$ and Leu$^+$ intragenic recombinants result almost exclusively from nonreciprocal gene conversion events (Hurst et al., 1972; Fink and Styles, 1974; Lambie and Roeder, 1988). In the *mer1* mutant, gene conversion is reduced to about 10 % of wild type (Table 1), a reduction similar to that observed for reciprocal crossing over.

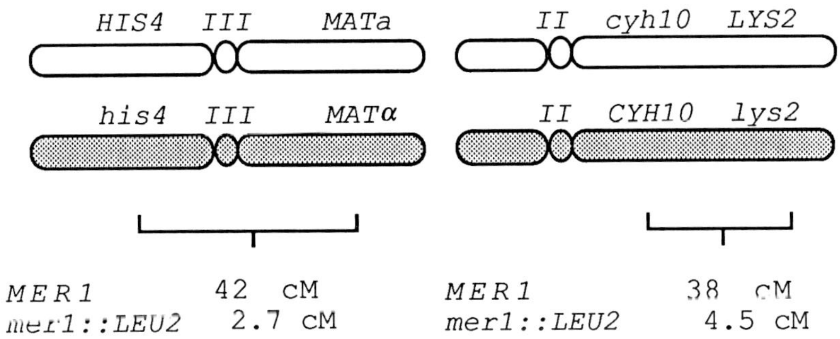

Figure 6. Map distances in wild type and *mer1*.

Intrachromosomal Recombination is Reduced in *mer1* Mutants

The 10-fold reduction in meiotic interchromosomal recombination observed in strains carrying *mer1* mutations is similar to the defect described for *hop1* mutants. Mutations at the *HOP1* locus do not affect intrachromosomal recombination between duplicated sequences on the same chromosome. To determine whether the *mer1* defect is also specific to interhomologue exchange, intrachromosomal recombination was measured.

The assay system described and constructed by Hollingsworth and Byers (1989) was used to measure recombination between duplicated sequences on chromosome III (Figure 7). One of the two copies of chromosome III contains a duplication of the

Table 1. Intragenic Recombination in *mer1* Mutants

Strain	Prototrophs	Type	Frequency	Fold
spo13				
	His Prototrophs	Mitotic	2.5×10^{-4}	
		Meiotic	9.1×10^{-3}	(36-fold)
	Leu Prototrophs	Mitotic	4.3×10^{-6}	
		Meiotic	1.3×10^{-3}	(302-fold)
mer1 spo13				
	His Prototrophs	Mitotic	4.0×10^{-4}	
		Meiotic	1.3×10^{-3}	(3.2-fold)
	Leu Prototrophs	Mitotic	3.4×10^{-6}	
		Meiotic	8.5×10^{-5}	(25-fold)
spo11 spo13				
	His Prototrophs	Mitotic	4.3×10^{-4}	
		Meiotic	4.2×10^{-4}	(0.9-fold)
	Leu Prototrophs	Mitotic	4.0×10^{-6}	
		Meiotic	3.8×10^{-6}	(0.9-fold)
mer1 spo11 spo13				
	His Prototrophs	Mitotic	2.7×10^{-4}	
		Meiotic	4.2×10^{-4}	(1.5-fold)
	Leu Prototrophs	Mitotic	2.9×10^{-6}	
		Meiotic	3.5×10^{-5}	(1.2-fold)
hop1 spo13				
	His Prototrophs	Mitotic	4.3×10^{-4}	
		Meiotic	1.8×10^{-3}	(4.1-fold)
	Leu Prototrophs	Mitotic	4.2×10^{-6}	
		Meiotic	1.2×10^{-4}	(29-fold)
mer1 hop1 spo13				
	His Prototrophs	Mitotic	6.6×10^{-4}	
		Meiotic	6.2×10^{-4}	(0.9-fold)
	Leu Prototrophs	Mitotic	3.6×10^{-6}	
		Meiotic	3.3×10^{-6}	(0.9-fold)

All strains are isogenic; the *mer1*, *spo11*, *hop1* and *spo13* mutations are deletion/disruption mutations. After the meiotic frequencies, the fold increase above mitotic background is indicated. *rad50 spo13* and *mer1 rad50 spo13* mutants gave results similar to those for *spo11 spo13* and *mer1 spo11 spo13* strains respectively (data not shown).

sequences between *LEU2* and *HIS4*. These repeats flank inserted vector DNA including the wild-type, cycloheximide-sensitive *CYH2* gene. Present at the *CYH2* locus on chromosome VII is a recessive, cycloheximide-resistant mutation (*cyh2*^{R}). Recombination between the repeats results in excision of the vector sequences, leading to loss of the *CYH2* marker and expression of the *cyh2*^{R} marker. In this intrachromosomal assay, the *mer1* mutant promotes only 8 % of wild-type levels of recombination. Thus, the *mer1* mutant is defective in exchange both between and within chromosomes.

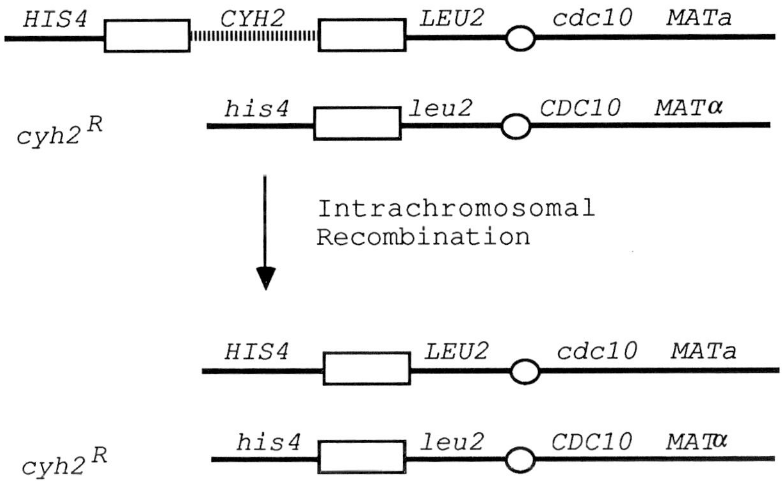

Figure 7. Assay for intrachromosomal recombination between repeated chromosome III sequences. Intrachromosomal recombinants are selected as His^{+}, Leu^{+} and Cyh^{R}. The pattern of chromosome segregation (equational versus reductional) must be taken into account in calculating recombination rates and can be monitored by following the centromere-linked, *cdc10*, mutation.

mer1 Rescues *rad52 spo13* Mutants

As discussed in the Introduction, *rad52 spo13* mutants produce inviable spores but this lethality can be rescued by a *spo11* or *rad50* mutation. To determine whether a *mer1* mutation can prevent *rad52*-induced lethality, *mer1 rad52 spo13* strains were constructed and analyzed for spore viability. Whereas a *rad52 spo13* strain produced 4 % viable spores, the *mer1 rad52 spo13* mutant produced 61 % viable progeny. This rescue of *rad52* by a *mer1* mutation suggests that the *MER1* gene product acts before the *RAD52* protein.

Recombination in Multiply Mutant *mer1* Strains

mer1 spo13 mutants display 10 % of wild-type levels of recombination; in contrast, *spo11 spo13* and *rad50 spo13* strains show no induction of meiotic exchange. To examine the epistatic relationships between *MER1* and *SPO11* (and *RAD50*), *mer1 spo11 spo13* and *mer1 rad50 spo13* mutants were constructed and analyzed for intragenic recombination. Meiotic recombination was not increased above mitotic background levels in either of the triple mutants (Table 1). The *spo11* and *rad50* mutations are therefore epistatic to *mer1* suggesting that the *SPO11* and *RAD50* proteins act before the *MER1* gene product.

Interchromosomal gene conversion was also measured in *hop1 mer1 spo13* strains. Whereas meiotic recombination in both *hop1 spo13* and *mer1 spo13* strains is increased 10-fold above mitotic background levels, the *mer1 hop1 spo13* triple mutant shows no meiotic conversion (Table 1). Thus, the defect in the triple mutant is more severe than that in each of the corresponding double mutants. This synergistic effect suggests that the *mer1* and *hop1* mutations affect different processes, both of which are essential for normal levels of meiotic recombination.

Role of the *MER1* Gene Product

The *MER1* gene product is required for normal levels of recombination within and between chromosomes. The phenotype of the *mer1 hop1 spo13* mutant suggests that *MER1* and *HOP1* are involved in different structures or functions, both of which are necessary for exchange. Antibodies to the *HOP1* gene product have localized this protein to the SC (L. Goetsch, N. Hollingsworth and B. Byers, personal communication), a result consistent with the hypothesis that *hop1* mutants are defective in the pairing of homologous chromosomes. The recombination machinery appears to be intact in *hop1* mutants since intrachromosomal recombination occurs at wild-type levels. In contrast, *mer1* mutants are reduced for both intra- and interchromosomal recombination. Taken together, these results suggest that the *MER1* gene product may be a component of the meiotic recombination apparatus.

If the *MER1* protein is a recombination enzyme, it probably catalyzes an early step in recombination, such as initiation. This is suggested by the ability of *mer1* mutations to rescue *rad52*-induced spore lethality. The fact that *mer1* mutants display some meiotically-induced recombination suggests the existence of a *MER1*-independent recombination pathway. This second pathway could employ other meiotically-induced enzymes; alternatively, paired chromosomes may serve as better substrates for mitotic enzymes present during meiosis.

To further explore the function of the *MER1* gene product, the *mer1* mutant will be assayed for recombination competence *in vitro* in cell-free extracts of meiotic yeast cells using the assay system described by Symington et al. (1984). If the *mer1* mutant is defective in recombination *in vitro*, then attempts will be made to complement this defect using *MER1* protein isolated from *E. coli* cells expressing the *MER1* gene. *In vitro* complementation would permit purification of the *MER1* gene product; the enzymatic activities catalyzed by the purified protein could then be characterized.

Screen for Recombination-Defective Meiotic Mutants in a *spo13* Strain Disomic for Chromosome III

In addition to the screen for meiotic-lethal mutants, we have used a second method to search for mutants defective in meiotic recombination. This procedure employs a haploid *spo13* strain disomic for chromosome III. Present on chromosome III are *MATa* and *MATα* information, as well as *his4* and *leu2* heteroalleles. Because this haploid strain carries a *spo13* mutation and expresses both *MATa* and *MATα* information, it is able to undergo meiosis to produce dyads containing viable 1n + 1 spores. The frequencies of His^+ and Leu^+ prototrophs provide measures of interchromosomal gene conversion. Recessive mutations can be easily recovered because the starting strain carries a single copy of all chromosomes except chromosome III. This scheme is similar to that described by Roth and Fogel (1971) except for the presence of the *spo13* mutation.

The *spo13* disomic strain was exposed to UV at a dose sufficient to obtain 50 % killing and the survivors were allowed to form colonies. These were patched onto rich medium and later replica plated to sporulation medium to induce meiosis. After several days, the sporulated patches of cells were transferred to medium lacking histidine and medium lacking leucine to assay interchromosomal recombination. Mutants that generated His^+ and Leu^+ prototrophs at reduced frequencies were examined for sporulation in the light microscope.

After screening 10,000 survivors of UV mutagenesis, seven sporulation-competent mutants that displayed reduced rates of meiotic recombination were identified. Complementation tests indicated that one of these is an allele of *RAD57*; the other six represent previously undescribed genes. Two are alleles of a gene designated *MEI4*, located on chromosome V.

The *mei4-1* Mutant is Rescued by *spo13*

Dyad dissection of the original *mei4-1* mutant isolate indicates that the *mei4-1 spo13* strain produces viable meiotic products. Thus, the failure to recover His^+ and Leu^+ meiotic recombinants is not a consequence of the production of inviable progeny. The meiotic products of the *mei4-1 spo13* mutant must therefore be nonrecombinant.

Through a series of genetic crosses, diploids homozygous for the *mei4-1* mutation and either *spo13* or *SPO13* were constructed. The *mei4-1 spo13* diploids produced 88 % viable spores whereas the *mei4-1 SPO13* strains produced only 5 % viable meiotic products. Thus, *mei4-1* is a meiotic lethal mutant which is rescued by a *spo13* mutation.

Interchromosomal Recombination is Eliminated in the *mei4-1* Mutant

The frequencies of mitotic and meiotic convertants in the *mei4-1* mutant were quantitated by physically isolating spores (Lambie and Roeder, 1988) and titering for His^+ and Leu^+ recombinants. Mitotic recombination occurred at wild-type levels in the *mei4-1* mutant. When the wild-type control strain was induced to sporulate,the frequencies of His^+ and Leu^+ prototrophs were increased 65- and 400-fold, respectively. In contrast, the *mei4-1* mutant showed no induction above mitotic background.

The effect of the *mei4-1* mutation on crossing over in the *LEU2-MAT* interval was examined in diploids homozygous for the *mei4-1* and *spo13* mutations. The *LEU2-MAT* interval measured 29 cM in wild type but only 1.5 cM in the *mei4-1* mutant. Thus, the *mei4-1* mutant is defective in both gene conversion and crossing over.

The recombination deficiency observed in *mei4* mutants may result from a failure to pair homologous chromosomes and/or a defect in the meiotic recombination machinery. Alternatively, the *MEI4* function may be a necessary prerequisite to the initiation of pairing and exchange. Future studies of the *mei4* mutant will include electron-microscopic observations of SC and assays of recombination *in vitro*.

ACKNOWLEDGMENTS

This research was supported by National Science Foundation grant PCM-8351607, National Institutes of Health grant GM28904 and a gift from the Du Pont Co..

REFERENCES

Adzuma K, Ogawa T, Ogawa H (1984). Primary structure of the *RAD52* gene in *Saccharomyces cerevisiae*. Mol Cell Biol 4:2735-2744.

Atcheson CL, DiDomenico B, Frackman S, Esposito RE (1987). Isolation, DNA sequence, and regulation of a meiosis-specific eukaryotic recombination gene. Proc Natl Acad Sci USA 84:8035-8039.

Baum P, Furlong C, Byers B (1986). Yeast gene required for spindle pole body duplication: homology of its products with Ca^{2+}-binding proteins. Proc Natl Acad Sci USA 83:5512-5516.

Borts RH, Lichten M, Haber JE (1986). Analysis of meiosis-defective mutations in yeast by physical monitoring of recombination. Genetics 113:551-567.

Dawes IW, Hardie ID (1974). Selective killing of vegetative cells in sporulated yeast cultures by exposure to diethyl ether. Mol Gen Genet 131:281-289.

Engebrecht J, Roeder GS (1989). Yeast *mer1* mutants display reduced levels of meiotic recombination. Genetics: in press.

Esposito MS, Esposito RE (1969). The genetic control of sporulation in *Saccharomyces cerevisiae*. I. The isolation of temperature-sensitive sporulation-defective mutants. Genetics 61:79-89.

Esposito RE, Esposito MS (1974) Genetic recombination and commitment to meiosis in *Saccharomyces*. Proc Natl Acad Sci USA 71:3172-3176.

Farnet C, Padmore R, Cao L, Raymond W, Alani E, Kleckner N (1989). The *RAD50* gene of *Saccharomyces cerevisiae*. In Friedberg E, Hanawalt P (eds): "Mechanisms and Consequences of DNA Damage Processing", New York: Alan R Liss, in press.

Fink GR, Styles CA (1974). Gene conversion of deletions in the *HIS4* region of yeast. Genetics 77:231-244.

Game JC, Zamb TJ, Braun RJ, Resnick M, Roth RM (1980). The role of radiation (*rad*) genes in meiotic recombination in yeast. Genetics 94:51-68.

Garber AT, Segall J (1986). The *SPS4* gene of *Saccharomyces cerevisiae* encodes a major sporulation-specific mRNA. Mol Cell Biol 6:4478-4485.

Giroux CN (1989). Chromosome synapsis and meiotic recombination. In Kucherlapati R, Smith GR (eds): "Genetic Recombination", American Society for Microbiology, in press.

Hollingsworth N, Byers B (1989). *HOP1*: a yeast meiotic pairing gene. Genetics: in press.

Hurst D, Fogel S, Mortimer RK (1972). Conversion-associated recombination in yeast. Proc Natl Acad Sci USA 69:101-1105.

Klapholz S, Esposito RE (1980). Recombination and chromosome segregation during the single division meiosis in *spo12-1* and *spo13-1* diploids. Genetics 96:589-611.

Klapholz S, Waddell CS, Esposito, RE (1985). The role of the *SPO11* gene in meiotic recombination in yeast. Genetics 110:187-216.

Lambie EJ, Roeder GS (1988) A yeast centromere acts in *cis* to inhibit meiotic gene conversion of adjacent sequences. Cell 52:863-873.

Malone RE (1983). Multiple mutant analysis of recombination in yeast. Mol Gen Genet 189: 405-412.

Malone RE, Esposito RE (1981). Recombinationless meiosis in *Saccharomyces cerevisiae*. Mol Cell Biol 1:891-901.

McLeod M, Stein M, Beach D (1987). The product of the *mei3*$^{+}$ gene, expressed under control of the mating-type locus, induces meiosis and sporulation in fission yeast. EMBO J 6:729-736.
Nasmyth, K A (1982). Molecular genetics of yeast mating type. Annu Rev Genet 16:439-500.
Percival-Smith A, Segall J (1986). Characterization and mutational analysis of a cluster of three genes expressed preferentially during sporulation of *Saccharomyces cerevisiae*. Mol Cell Biol 6:2443-2451.
Resnick MA, Nitiss J, Edwards C, Malone RE (1986). Meiosis can induce recombination in *rad52* mutants of *Saccharomyces cerevisiae*. Genetics 113:531-550.
Rockmill B, Fogel S (1988). *DIS1*: a yeast gene required for proper meiotic chromosome disjunction. Genetics 119: 261-272.
Rockmill B, Roeder GS (1988). *RED1:* A yeast gene required for the segregation of chromosomes during the reductional division of meiosis. Proc Natl Acad Sci USA 85:6057-6061.
Roth R, Fogel S (1971). A system selective for yeast mutants deficient in meiotic recombination. Mol Gen Genet 112:295-305.
Seifert HS, Chen EY, So M, Heffron F (1986). Shuttle mutagenesis: a method of transposon mutagenesis for *Saccharomyces cerevisiae*. Proc Natl Acad Sci USA 83: 735-739.
Symington LS, Morrison PT, Kolodner R (1984). Genetic recombination catalyzed by cell-free extracts of *Saccharomyces cerevisiae*. Cold Spring Harbor Symp Quant Biol 49:805-814.
Thompson EA, Roeder GS. Expression and DNA sequence of *RED1*, a gene required for meiosis I chromosome segregation in yeast. Submitted.
Wagstaff JE, Klapholz S, Esposito RE (1982). Meiosis in haploid yeast. Proc Natl Acad Sci USA 79:2986-2990.
Wang H, Frackman S, Kowalisyn J, Esposito RE, Elder R (1987). Developmental regulation of *SPO13,* a gene required for separation of homologous chromosomes at meiosis I. Mol Cell Biol 7:1425-1435.

Molecular and Cytogenetic Studies of
Non-Disjunction, pages 327–348

MEIOTIC SEGREGATION OF NORMAL AND DELETION CHROMOSOMES IN *SACCHAROMYCES CEREVISIAE*

Richard T. Surosky and Bik-Kwoon Tye

Department of Biochemistry, Cell and Molecular Biology, Cornell University, Ithaca, New York 14853

ABSTRACT

We explored the behavior of meiotic chromosomes in *Saccharomyces cerevisiae* by examining the effects of chromosomal rearrangements on recombination and disjunction. Chromosome III derivatives in which the entire left arm or the entire right arm was deleted (telocentric) segregated with fidelity from a normal chromosome III. Recombination between either of these two deletion chromosomes and the intact chromosome also appeared normal. In a strain containing a right arm telocentric, a left arm telocentric and one normal chromosome both telocentrics disjoined from the normal chromosome. Homology on one arm was sufficient for proper recombination and segregation of these chromosomes. In strains containing two normal chromosomes and one telocentric chromosome the two normal chromosomes preferentially disjoined. In a few cases however, the two normal chromosomes cosegregated. Recombination between the two normal chromosomes or between one normal chromosome and the deletion chromosome increased the probability that they would disjoin, although cosegregation of recombinants was observed. A chromosome III derivative which contained a large centromeric deletion and an insertion of the centromere from chromosome V into a nonhomologous position segregated with fidelity from a normal chromosome III. These studies demonstrate that it is not pairing of the centromeres, but pairing and recombination along the arms of the homologs that directs meiotic chromosome disjunction.

INTRODUCTION

In meiosis, chromosomes undergo two successive segregations, reducing the chromosome number by half. Proper chromosome segregation is essential to insure that aneuploid cells, which may be aberrant or inviable are not produced. In *Saccharomyces cerevisiae*, meiotic chromosome segregation is part of a process termed sporulation (reviewed by Esposito and Klapholz, 1981). Diploid yeast cells, when placed in the appropriate medium, undergo meiosis and spore formation. Biochemical and cytological analyses of meiosis in *S. cerevisiae* indicate that it is similar to meiosis in other eukaryotes (Hooper et al., 1974; Zickler and Olson, 1975).

Several approaches have been utilized to examine meiotic chromosome segregation. Cytological analysis has demonstrated that an orderly series of events leading to chromosome disjunction occurs in nearly all organisms. Chromosome pairing, which frequently is initiated at the ends of the chromosomes, is mediated by the formation of the synaptonemal complex between the homologs (Moens, 1969a). The formation of this complex, which holds the two homologs in close proximity, is required for recombination to occur in most organisms (reviewed by Gillies, 1975; Westergaard and von Wettstein, 1972). Meiotic recombination events are visualized as chiasma, physical attachments between the homologs (Mather, 1933; White, 1954). After breakdown of the synaptonemal complex, the bivalents, held together only at the chiasma, attach to the spindle poles through the spindle fibers (Westergaard and von Wettenstein, 1970). Tension on the bivalent generated by the attachment of one kinetochore to one spindle pole and the homologous kinetochore to the other spindle pole stabilizes this arrangement and facilitates segregation (Nicklas and Koch, 1969; Nicklas, 1974). If both homologs become attached to the same spindle pole, reorientation will occur until opposing tension is generated (Hughes-Schrader, 1943). The chromosomes then disjoin, with one set of chromosomes segregating to each pole.

The analysis of meiotic mutants has also indicated a relationship between pairing, recombination and segregation. In Drosophila, several mutations that interefere with chromosome pairing or recombination have been identified and almost all increase the frequency of meiotic nondisjunction (reviewed by Baker and Hall, 1976). In yeast, the *spo11* and *rad50* mutations, which prevent meiotic recombination, lead to the production of inviable aneuploid spores (Klapholz et al., 1985; Game et al., 1980). Spores from a *spo11* mutant can be rescued by the introduction of a second mutation that causes a bypass of the first meiotic division (Malone and Esposito, 1981).

We have examined the meiotic behavior of a set of deletion derivatives of a natural chromosome in *S. cerevisiae* as another approach to the analysis of meiotic chromosome disjunction. We have constructed strains containing one or two copies of a normal chromosome III and derivatives in which all or part of one arm of this chromosome has been deleted. The meiotic recombination and segregation patterns were then analyzed and in most strains the deletion derivatives behaved like a normal chromosome. We also observed a correlation between meiotic recombination and chromosome segregation.

RESULTS

The construction of deletion derivatives of chromosome III has previously been described (Surosky and Tye, 1985a; Surosky et al., 1986). The derivatives used in all but the last experiment contain a deletion extending from the repetative telomeric sequence *Y'* to some internal site on the chromosome (Fig. 1). Chromosome dPT contains a deletion between *Y'* and the *PGK1* locus on the right arm. In the dLT chromosome, sequences between *Y'* and the *LEU2* gene on the left arm have been deleted. In chromosomes dC_RT and dC_LT, all sequences between *Y'* and the centromere on the right arm or the left arm, respectively, have been deleted. The dC_RT and dC_LT chromosomes are referred to as telocentric, although they contain approximately 10 kb of DNA between the centromere and the end of the chromosome on the deletion arm. The left arm deletions on chromosomes dLT and dC_LT are marked by the insertion of the *URA3* gene and the right arm deletions on chromosomes dPT and dC_RT are marked by the insertion of the *LEU2* gene. In the last cross a chromosome containing a 66 kb deletion between the *LEU2* region and the *PGK1* region including the centromere was examined. In addition, the centromere from chromosome V (*CEN5*) was integrated at the *HIS4* locus (Fig. 1) (Surosky and Tye, 1985b). Previous studies have demonstrated that these chromosomes are mitotically stable (Surosky and Tye, 1985a; Surosky et al., 1986), so it is unlikely that any aberrant meiotic segregation that is observed results from mitotic loss of these chromosomes.

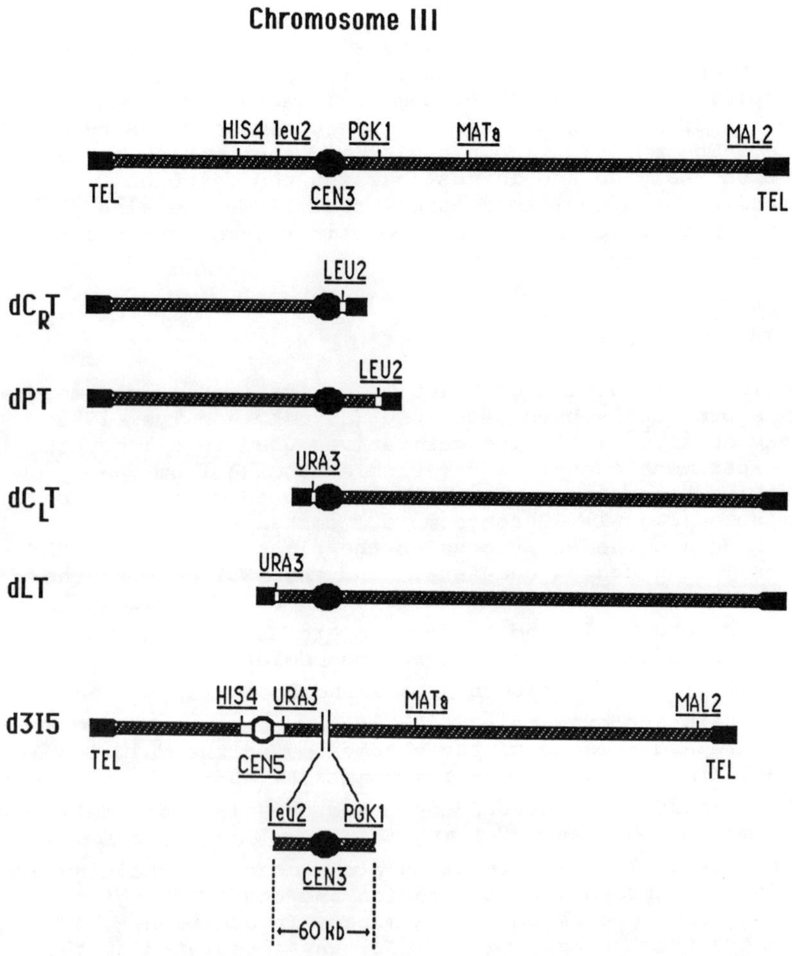

Figure 1. Deletion chromosomes used in this study.

TABLE 1. Yeast Strains Used in this Study

Strain	Genotype	Ploidy and Chromosome Rearrangements
8534-8C	*MATα his4 leu2 ura3*	
320-13B	*MAT**a** ura3 met3 trp1 MAL2*	
320-16C	*MAT**a** leu2 ura3 met3 trp1 MAL2*	
DLT1	*his4 leu2 MAT**a** mal2 ura3 MET3 TRP1* / *[URA3] LEU2 MATα MAL2 ura3 met3 trp1*	Diploid with one normal III and one dLT chromosome.
DC_LT	*his4 leu2 MAT**a** mal2 ura3 MET3 TRP1* / *[URA3] MATα MAL2 ura3 met3 trp1*	Diploid with one normal III and one dC_LT chromosome.
DPT	*his4 leu2 MATα mal2* / *HIS4 leu2 [LEU2]* *ura3::[URA3 MAT**a**] MET3 TRP1* / *ura3 met3 trp1*	Diploid with one normal chromosome III and one dPT chromosome.

(Continued)

TABLE 1, continued

Strain	Genotype	Comments
DC_RT	*his4 leu2 MATα mal2 ura3::[URA3 MAT**a**]* / *HIS4 leu2 [LEU2] ura3* *MET3 TRP1* / *met3 trp1*	Diploid with one normal chromosome III and one dC_RT chromosome.
338-15A	*his4 leu2 MATα ura3 met3* / *HIS4 leu2 [LEU2]*	Strain contains one normal chromosome III and one dC_RT chr.
338-115B	*his4 leu2 MATα ura3 met3 trp1* / *HIS4 leu2 [LEU2]*	Strain contains one normal chromosome III and one dC_RT chr.
RS2-145	*his4 leu2 MATα ura3 met3 trp1* / *HIS4 leu2 [LEU2]*	Strain contains one normal III and one dPT chr.
RS3-3	*leu2 MATα ura3 met3 trp1* / *leu2 [LEU2]*	Strain contains one normal chromosome III and one dC_RT chr.
AH2	*MAT**a** his4 leu2*	
AGH5	*MATα leu2 ura3*	

TABLE 1, continued

425-6D	*MAT***a** *leu2 his4 ura3 met3 MAL2*	
426-67A	*MAT***a** *leu2 his4*::[*URA3*] *ura3 MAL2*	*URA3* gene replaced *his4* gene.
353-14A	*his4 leu2* [*URA3*] *ura3 met3* / [*LEU2*] *MAT***a** *MAL2*	Strain contains dC_RT and dC_LT chromosome.
353-14ALU	*MAT***a** *his4 leu2 ura3 met3 trp1 MAL2*	Mitotic recombination between dC_RT and dC_LT chr.
ICH2	*MAT***a** *his4*::[*HIS4 URA3*] *trp1*	Strain contains ring and linear chromosome in place of chr. III.

Deletion Chromosomes Segregate with Fidelity.

In the first set of experiments we constructed diploids containing one normal chromosome III and one deletion chromosome and examined the meiotic segregation. Strains DLT and DC_LT contain the left arm deletion chromosomes dLT and dC_LT, respectively. Strains DPT and DC_RT contain the right arm deletion chromosomes dPT and dC_RT, respectively. Due to the extensive deletion on one chromosome, these diploids produce no more than two viable spores. If the normal and deletion chromosomes disjoin at meiosis I, then the two viable spores would contain only the normal chromosome and therefore be Ura^- or Leu^-. If the normal and deletion chromosomes cosegregate, then pairs of Ura^+ or Leu^+ spores would be produced. In diploids DPT and DC_RT the *MAT**a*** locus has been deleted. A plasmid containing the *MAT**a*** gene was integrated at *ura3* to sporulate these diploids.

TABLE 2. Crosses Used in this Study

Cross	Diploid	Parents
1	DLT	8534-8C x 320-13B
2	DC_LT	8534-8C x 320-13B
3	DPT	8534-8C x 320-16C
4	DC_RT	8534-8C x 320-16C
5		353-14A x AGH5
6		353-14ALU x AGH5
7		338-115B x AH2
8		RS3-3 x AH2
9		338-15A x 426-67A
10		RS2-145 x AH2
11		RS2-145 x 426-67A
12		ICH2 x 425-6D

Over two hundred tetrads were analyzed for each diploid and at least 80% of the tetrads yielded two viable spores (Fig. 2). Only the DC_RT diploid produced pairs of Leu^+ spores

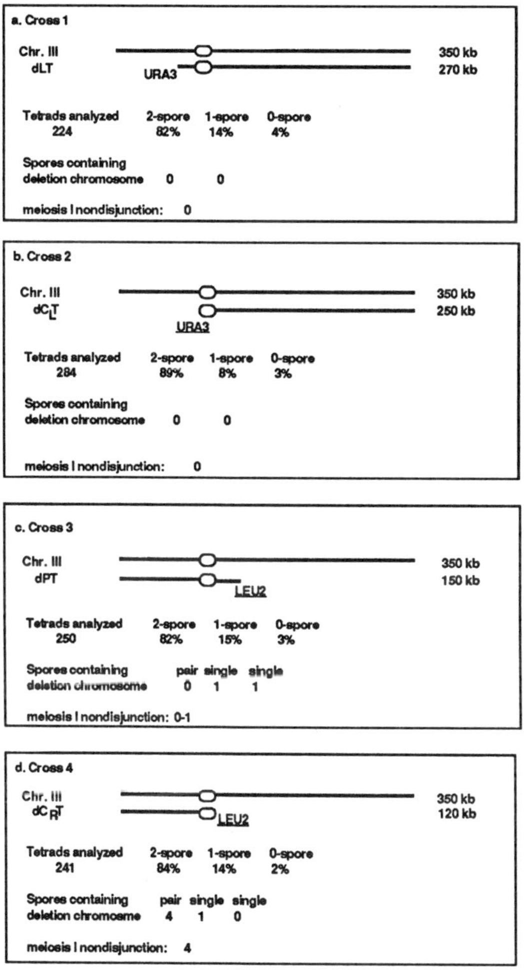

Figure 2. Chromosome segregation in strains containing one normal chromosome III and one deletion chromosome. Spores were initially identified as Ura^+ or Leu^+ by replica plating onto Cm-leu or Cm-ura media. The presence of the deletion chromosome in the Leu^+ or Ura^+ spores was verified by hybridization and genetic analysis. In the figure, 2-, 1- and 0-spore indicate the number of viable spores per tetrad. Pair and single indicate the number of spores per tetrad containing the deletion chromosome.

which upon further analysis were found to contain one normal and one deletion chromosome. Simple nondisjunction occurred at a frequency of 2% in the DC_RT strain and was undetectable in the other diploids. A few single spores containing one normal and one deletion chromosome were also recovered (Fig. 2). These spores may have resulted from other types of aberrant segregation such as premature sister chromatid disjunction. Clearly in these diploids the normal and deletion chromosomes segregated with fidelity and rarely was nondisjunction observed.

The levels of meiotic recombination between the normal and deletion chromosomes were also determined. Intervals on both the deletion arm for the dPT and dLT chromosomes and the intact arm for all chromosomes were examined. Although no more than two viable spores were obtained, the centromere linkage of chromosome III markers could be measured with respect to *trp1* and *met3*, two tightly centromere-linked markers on other chromosomes. In comparison to the parental strains containing two intact chromosome IIIs, all intervals showed similar or higher rates of recombination. These results support those from the segregation analysis, that there is no impairment of normal meiotic behavior due to a deletion on one arm of chromosome III.

Segregation Pattern of One Normal Chromosome and Two Telocentric Chromosomes Containing Opposite Arms.

After determining that one telocentric and one normal chromosome segregate properly, a strain containing one normal chromosome and one right arm and left arm telocentric chromosome was constructed. Disjunction of both telocentric chromosomes from the normal homolog would produce four viable spores, two Leu^+Ura^+ and two Leu^-Ura^-. Other segregation patterns would yield only two viable spores with various Leu and Ura phenotypes.

Tetrads from this strain were analyzed in cross 5 (Fig. 3a). A high percentage of four-viable-spore tetrads (60.0%) as well as an unexpectedly high percentage of three-viable-spore tetrads (20.9%) were obtained. In cross 6 we repeated this analysis using a strain isogenic to the strain in cross 5 only containing two normal chromosome IIIs (Fig. 3b). Again a high percentage of three-viable-spore tetrads was obtained suggesting that these tetrads did not result from the presence of the telocentric chromosomes, but are due to some other property of the strain. If the three- and four-viable-

a. Cross 5

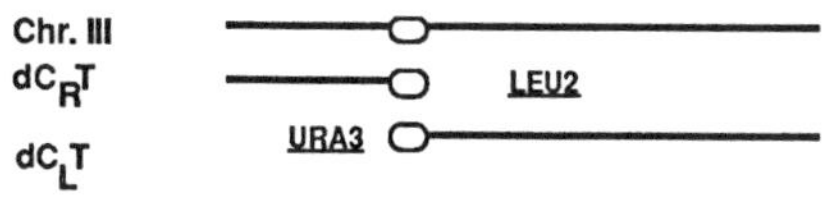

	spores	% Viable	Class	%
345 tetrads analyzed:	4	60.0	2L+U+/2L-U-	58
	3	20.8		
	2	15.6	2L+U- 2L-U+	6.4 2.3
	1	1.4		
	0	2.0		

b. Cross 6

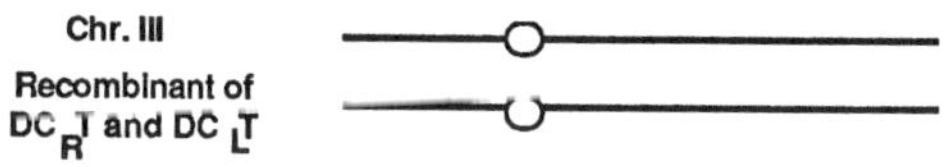

279 tetrads analyzed:		
	4-spore	61%
	3-spore	19%
	2-spore	13%
	1-spore	3.2%
	0-spore	3.6%

Figure 3. a) Analysis of tetrads derived from a strain containing one normal chromosome III, one right arm telocentric and one left arm telocentric chromosome. b) Isogenic strain in which the two telocentric chromosomes have recombined to regenerate a normal chromosome.

sprore tetrads are combined then it appears that the two telocentrics disjoined from the normal chromosome in over 80% of the first meiotic divisions. The high percentage of two Leu^+Ura^+, two Leu^-Ura^- spores among the four-viable-spore tetrads verifies this pattern of segregation. The tetrads producing two viable spores were also further analyzed. Over half of the tetrads in this class resulted from cosegregation of one of the telocentric chromosomes with the normal chromosome.

Chromosome Segregation in Diploids Containing Two Normal Chromosomes and One Deletion Chromosome.

We determined if the deletion chromosomes could be distinguished from an intact chromosome in meiotic segregation by constructing strains containing two normal chromosome IIIs and one left arm deletion chromosome. If the deletion chromosome is equivalent to the normal chromosome, any two chromosomes will have an equal probability of disjunction and one-third of the tetrads produced by this strain will contain only two viable spores both of which are Leu^-. In contrast, if the two normal chromsomes always disjoin then all tetrads will contain four viable spores.

The tetrads obtained from these strains contain more four viable spores and less two viable spores than would be expected if there was an equal probability of disjunction between any two chromosomes (Fig. 4). The actual frequency of nondisjunction of the two normal chromosomes was determined by analyzing tetrads containing two viable spores, both of which were Leu^-. Tetrads of this type were obtained at a frequency of 5.2% from the strain containing the left arm telocentric chromosome (Fig. 4a). Some of these spores were mated or fused to tester strains containing multiple markers on chromosome III to determine whether the spores were disomic for chromosome III. Adjusting our values for these results we obtain a frequency of 4.2% for normal chromosome III nondisjunction. In tetrads derived from strains containing two normal chromosomes and the dPT chromosome, the frequency of nondisjunction of the two normal chromosomes is 7.1% (Fig 4b). This slightly higher frequency may have resulted from the addition of 20 kb of homologous sequence not present on the dC_RT chromosome. Both of these frequencies are well below the 33% expected if random segregation of the chromosomes occurred.

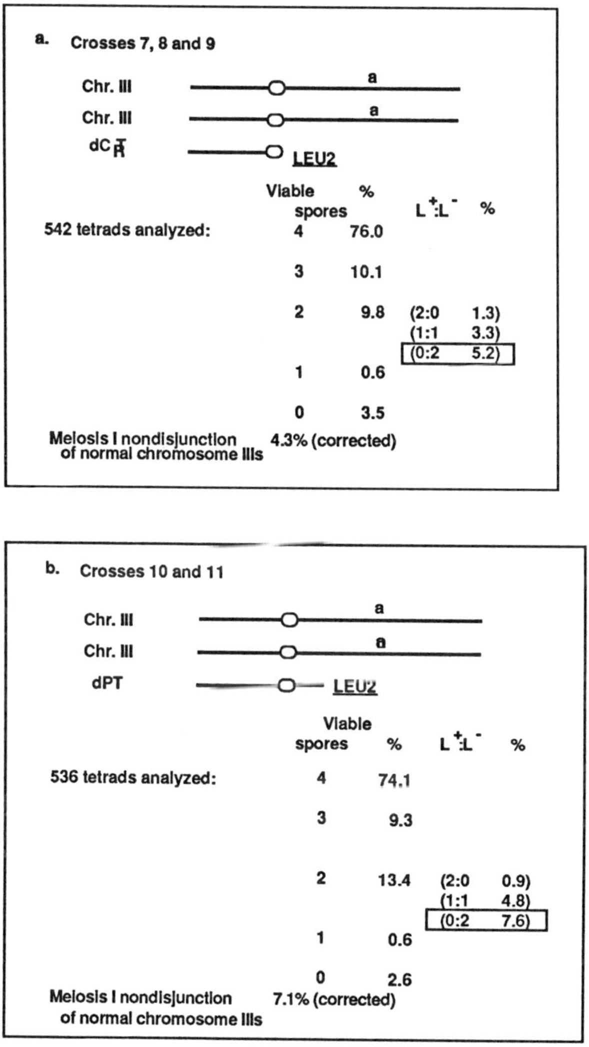

Figure 4. Analysis of tetrads derived from strains containing two normal and one deletion chromosome. The percentages of spore viability and *leu2* segregation are calculated from the total tetrads analyzed.

Relationship Between Segregation and Recombination.

In Crosses 9 and 11, recombination on chromosome III was examined. First, recombination between the normal chromosomes in the *MAT-cen3* interval on the right arm of the chromosome was analyzed. Nonrecombinants in this interval were more frequent among normal chromosomes that did not disjoin (81.2%) than among the chromososmes that did disjoin (46.2%), suggesting that recombination between the two normal chromosomes on the right arm increased the probability that they would disjoin. Recombination between the two normal chromosomes however, did not insure their disjunction . Among the pairs of spores in which the two normal chromosomes cosegregate spores that contain two *MAT**a*** or two *MATα* alleles, which most likely result from recombination in the *MAT-cen3* interval, were recovered.

To further examine this relationship, recombination on the left arm of the chromosome in the *his4-cen3* interval was examined. These strain were constructed such that the left arm of the deletion chromosome is marked by *HIS4*, the left arm of one normal chromosome is marked by *his4* and the left arm of the other normal chromosome is marked by *URA3*. Thus, recombination between any two chromosomes can be detected as a second division segregation event of the markers associated with those chromosomes (Table 3). Among the tetrads in which the normal chromosomes disjoined, most are either nonrecombinant in this interval (type 1A) or contain recombinants between the two normal chromosomes (type 2A). In the smallest class, a recombination event between the deletion chromosome and one normal chromosome had occurred. Among this latter class, in most cases the deletion chromosome disjoined from the normal chromosome it had recombined with (type 3A versus 4A). This suggests that even in the cases in which the two normal chromosomes disjoin, the segregation of the deletion chromosome is directed by recombination.

Among the tetrads in which the two normal chromosomes cosegregated, the frequency that some classes were obtained was altered dramatically, although it should be noted that the sample size is much smaller. In a much higher percentage of tetrads, a recombination event had occurred between the deletion chromosome and one normal chromosome (type 3B versus 3A). In none of the tetrads in which the two normal chromosomes cosegregated was recombination between the normal chromosomes observed (type 2B). It appears that recombination between the normal chromosomes in the *his3-cen3* interval insured their disjunction while recombination between the

TABLE 3. Relationship Between Recombination and Segregation

	Class 1		Class 2		Class 3		Class 4	
	Nonrecombinant		Recombination between two normal chromosome IIIs		Recombination between deletion chromosome and one normal chromosome III -- they disjoin		Recombination between deletion chromosome and one normal chromosome III -- they cosegregate	
	A Normal Chromosome III disjunction[a]	B Normal Chromosome III nondisjunction	A Normal Chromosome III disjunction	B Normal Chromosome III nondisjunction[b]	A Normal Chromosome III disjunction	B Normal Chromosome III nondisjunction	A Normal Chromosome III disjunction[c]	B Normal Chromosome III nondisjunction
Cross 9	38.0-39.7% (72-75/189)	58.3% (7/12)	39.1% (74/189)	0 (0/12)	17.9% (34/189)	41.6% (5/12)	0-3.2% (0-6/189)	0
Cross 11	42.8-46.7% (81-86/184)	11.1% (1/9)	36.4% (67/184)	0 (0/9)	13.0% (26/184)	88.8% (8/9)	1.1-5.4% (2-10/184)	0

Only tetrads in which no more than one recombination event occurred in this interval were scored. Tetrads with segregation patterns that may have resulted from multiple crossovers, may also have been produced by other types of events and, due to this ambiguity, were not scored.

[a] The range of percentages reflects the fact that some spores classified as type 1A may result from a class 4A event.

[b] Only one-half of the events of this type can be identified. The other half will produce spores with the same phenotype as in type 1B.

[c] This range results from the fact that only one-half of the events of this type can be identified. The other half will produce spores with the same phenotypes as those resulting from a type 1A event.

deletion chromosome and one normal chromosome in this interval increased the probability that the normal chromosomes would cosegregate.

Pairing of the Centromeric Region Is Not Required for Proper Meiotic Segregation.

Previous studies have suggested that homology at the centomere or homologous positioning of the centromeres is not required for proper meiotic disjunction (Clarke and Carbon, 1983; Lambie and Roeder, 1986). We extented these observations by examining the importance of pairing of the regions flanking the centromeres in meiotic chromosome segregation. A strain containing one normal chromosome and one derivative in which the centromeric region including *CEN3* was deleted and *CEN5* was integrated at the *HIS4* region was constructed (Fig. 1). When this chromosome paired with an intact chromosome, approximately 30 kb on either side of *CEN3* on the normal chromosome and 7 kb around *CEN5* on the deletion chromosome would be unpaired. Due to the deletion on chromosome d315, tetrads from a diploid containing this chromosome yielded no more than two viable spores. Pairs of $His^{+}Ura^{+}$ spores, indicative of cosegregation of the normal and deletion chromosome were not recovered (Fig. 5). A few single $His^{+}Ura^{+}$ were obtained and may have resulted from premature sister chromatid nondisjunction of the deletion or normal chromosome.

DISCUSSION

In this study the meiotic segregation and recombination of deletion derivatives of chromosome III were examined. From this analysis the following conclusions can be drawn: 1. A chromosome lacking all or part of one arm recombines with and disjoins from its intact homolog. 2. Two deletion chromosomes with nonoverlapping sequences can both disjoin from a normal homolog. 3. In strains containing two normal chromosomes and one deletion chromosome the two normal chromosomes will preferentially disjoin, although cosegregation of the two normal chromosomes is observed. 4. In a competitive situation, recombination between chromosomes increases the probability, but does not guarantee their disjunction. 5. Pairing of the region around the centromere is not required for proper disjunction.

Cross 12

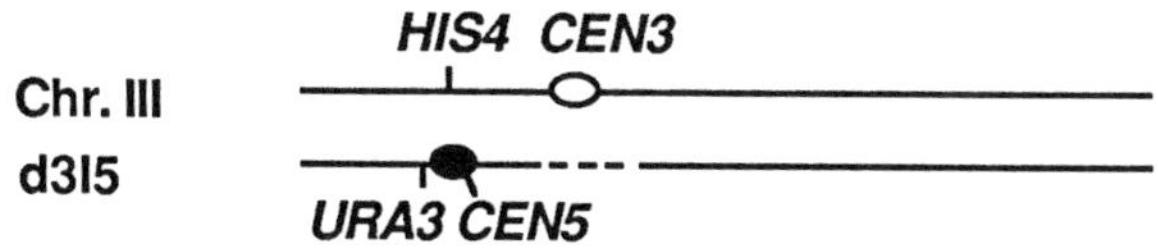

Tetrads analyzed	2-spore		1-spore	0-spore
226	76.1%		18.5%	5.3%
	pair	single	single	
His$^+$Ura$^+$ spores	0	4	1	
meiosis I nondisjunction:	0 - 1			

Figure 5. Analysis of tetrads derived from a strain containing one normal chromosome III and the centromeric deletion chromosome d315. Pair and single indicate the number of spores per tetrad that were His$^+$Ura$^+$.

In strains containing one normal chromosome III and one deletion derivative, proper segregation was observed in nearly all tetrads. Chromosomes with less than half of the normal amount of homology (dPT) or with no homolgy on one arm (dC_LT) segregated with fidelity. These results appear to rule out the existence of a unique site on the chromosome required for pairing or segregation unless it resides at the telomeres. Th possibility of many such sites spaced along the chromosome still exists. The 120 kb dC_RT chromosome displayed nondisjunction in 2% of the tetrads. In these cases the chromosomes may not have paired or may have paired, but not recombined due to decreases in the amount of homologous sequence. Alternatively, these tetrads may have resulted from an increased copy number of the deletion chromosome. Previous studies have demonstrated that decreases in chromosome size may yield increases in chromosome copy number (Murray and Szostak, 1983; Surosky et al., 1986).

A left and right arm telocentric chromosome simultaneously disjoined from an intact chromosome. Although these three chromosomes contain overlapping sequences at the centromere, the ability of each telocentric to pair and recombine noncompetitively with one arm of the normal chromosome was sufficient for disjunction. The results from this cross suggest that pairing of the centromeres is not required for proper segregation. This is also indicated by the segregation of chromosome d315 containing the large centromeric deletion. When this chromosome paires with its intact homolog, approximately 30 kb on each side of *CEN3* and 7 kb on each side of *CEN5* remain unpaired, yet these chromosomes segregate with fidelity.

The behavior of a deletion chromosome could be distinguished from a normal chromosome when a strain containing two normal and one deletion chromosome was examined. The two normal chromosomes preferentially disjoined, although cosegregation was observed in 4-7% of the tetrads. The deletion chromosome could not compete with a second normal chromosome. We believe that this can be explained by the recombination between the three chromosomes. Cytological studies have demonstrated that pairing of three homologous chromosomes may occur, although only two may be paired at any one position (Moens, 1969b).

The ability of the two normal chromosomes to pair and recombine on the right arm may play an important role in the disjunction of these two chromosomes. A recombination event in the *MAT-cen3* interval was observed to increase the probability that the two normal chromosomes would disjoin.

Recombination on the left arm in the *his4-cen3* interval also played an important role in disjunction. Recombination between the two normal chromosomes in this interval insured their segregation. When a recombination event was observed between the deletion chromosome and one normal chromosome the probability of nondisjunction of the normal chromosomes was increased. Interestingly, even in cases in which the two normal chromosomes disjoined, if the deletion chromosome recombined with one normal chromosome, it preferentially disjoined from that chromosome.

The behavior of the deletion chromosomes can be explained both in terms of centromere orientation and opposing tension on the homologs. The chiasma resulting from recombination orient the centromeres of the homologs toward opposite spindle poles increasing the probability they will become attached to opposite poles (Ostergren, 1951; Nicklas, 1967). At this time in meiosis the chromosomes must have sufficient rigidity so that once oriented by the chiasma, even on only one arm, this orientatation is maintained at the centromeres. The chiasma also generate an opposing tension on the homologs when the centromeres attach to opposite spindle poles. This tension stabilizes microtubule attachment and prevents reorientation (Nicklas and Koch, 1969; Nicklas, 1974).

The only requirement for proper disjunction is sufficient homology for efficient pairing and recombination. This criteria was met when any of the deletion chromosomes was combined with a single normal chromosome (Fig. 6a). Recombination of the right and left arm telocentric with the normal chromosome could also direct the disjunction of both telocentrics from the normal chromosome (Fig. 6b). The chiasma would orient the centromeres of both telocentrics in opposition to the centromere of the normal chromosome. Opposing tension would be generated by the attachment of the telocentrics to one spindle pole and the normal chromosome to the other. No homology at the centromeres or pairing of the centromeric region is required for disjunction (Fig. 6d).

When there is overlapping homology between three chromosomes the two chromosomes most likely to recombine are most likely to disjoin. Homology in the centromeric region may be especially important. In strains containing two normal and one deletion chromosome, recombination could direct both a normal and deletion chromosome to disjoin from a second normal chromosome (Fig. 6c). Recombination between the two normal chromosomes on the left and right arm locks their centromeres in opposition and insures their disjunction. However, when the deletion chromosome recombines with one of

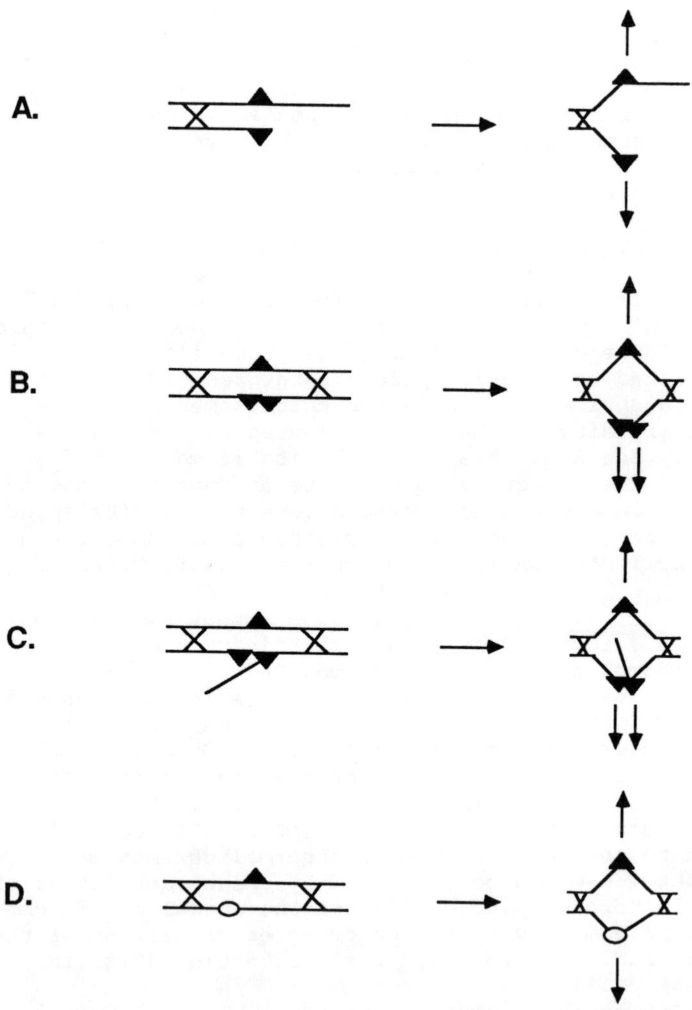

Figure 6. Pairing and recombination direct meiotic chromosome segregation. A) One normal and one deletion chromosome. B) One normal and two telocentric chromosomes. C) Two normal and one deletion chromosome. D) One normal and one centromeric deletion chromosome.

the normal chromosomes the orientation of the centromere of the other normal chromosome may be perturbed causing cosegregation of the two normal chromosomes.

As in previous studies, these experiments suggest an important role for recombination in directing chromosome segregation. Strains containing new deletion chromosomes or new combinations of normal and deletion chromosomes may be constructed to further examine these ideas.

This work was supported by Grant MV-233 from the American Cancer Society.

REFERENCES

Baker BS, Hall JC (1976). Meiotic mutants: Genic control of meiotic recombination and chromosome segregation. In Ashburner M, Novitski E (eds): "The Genetics and Biology of Drosophila" Vol. 1a, New York: Academic Press, pp. 352-434.

Clarke L, Carbon J (1983). Genomic substitutions of centromeres in *Saccharomyces cerevisiae*. Nature 305:23-28.

Esposito R, Klapholz S (1981). Meiosis and ascospore development. In Strathern J, Jones E, Broach J (eds): "The Molecular Biology of the Yeast Saccharomyces: Life Cycle and Inheritance," Cold Spring Harbor, New York: Cold Spring Harbor Laboratory, pp. 211-287..

Game J, Zamb TJ, Braun RJ, Resnick M, Roth RM (1980). The role of radiation (*rad*) genes in meiotic recombination in yeast. Genetics 94:51-68.

Gilles CB (1975). Synaptonemal complex and chromosome structure. Ann Rev Genet 9:91-109.

Hopper AK, Magee PT, Welch SK, Friedman M, Hall BD (1974). Macromolecule synthesis and breakdown in relation to sporulation and meiosis in yeast. J Bacteriol 119:619-628.

Hughes-Schrader S (1943). Polarization, kinetochore movement and bivalent structure in the meiosis of male mantids. Biol Bull 85:265-300.

Klapholz S, Waddell CS, Esposito RE (1985). The role of the *SPO11* gene in meiotic recombination in yeast. Genetics 110:187-216.

Lambie EJ, Roeder GS (1986). Repression of meiotic crossingover by a centromere (*CEN5*) in *Saccharomyces cerevisiae*. Genetics 114:769-789.

Malone RE, Esposito RE (1981). Recombinationless meiosis in *Saccharomyces cerevisiae*. Mol Cell Biol 1:891-901.

Mather K (1933). Interlocking as a demonstration of the occurrence of genetical crossing-over during chiasma formation. Amer Nat 67:476-479.

Moens PB (1969a). The fine structure of meiotic chromosome polarization and pairing in *Locusta migratoria* spermatocytes. Chromosoma 28:1-2.
Moens PB (1969b). Fine structure of meiotic chromosome pairing in the triploid, *Lilium tigrinium*. J Cell Biol 40:273-279.
Murray AW, Szostak JW (1983). Construction of artificial chromosomes in yeast. Nature 305:189-193.
Moses MJ (1968). Synaptonemal complex. Ann Rev Genet 2:363-412.
Nicklas RB (1967). Chromosome micromanipulation II. Induced reorientation and the experimental control of segregation in meiosis. Chromosoma 21:17-50.
Nicklas RB (1974). Chromosome segregation mechanism. Genetics 78:205-213.
Nicklas RB, Koch CA (1969). Chromosome micromanipulation III. Spindle fiber tension and the reorientation of mal-oriented chromosomes. J Cell Biol 43:40-50.
Ostergren G (1951). The mechanism of co-orientation in bivalents and multivalents. Heredetis 37:85-156.
Surosky RT, Tye B-K (1985a). Construction of telocentric chromosomes in *S. cerevisiae*. Proc Natl Acad Sci USA 82:2106-2110.
Surosky RT, Tye B-K (1985a). Resolution of dicentric chromosomes by Ty-mediated recombination in yeast. Genetics 110:397-419.
Surosky RT, Newlon CS, Tye B-K (1986) The mitotic stability of deletion derivatives of chromosome III in yeast. Proc Natl Acad Sci USA 83:414-418.
Westergaard M, von Wettstein D (1970). Studies on the mechanism of crossing over IV. The molecular organization of the synaptinemal complex in *Neottiella* (Cooke) Saccardo (Ascomyates). C R Trav Lab Carlsberg 37:239-268.
Westergaard M, von Wettstein D (1972). The synaptinemal complex. Ann Rev Genet 6:71-110.
White MJD (1954). "Animal Cytology and Evolution." Cambridge, Great Britain: University Press.
Zickler D, Olson LW (1975). The synaptonemal complex and the spindle plaque during meiosis in yeast. Chromosoma 50:1-23.

Index